Terrorism Handbook for Operational Responders

Armando Bevelacqua
Richard Stilp

Delmar Publishers

an International Thomson Publishing company I(T)P®

Albany • Bonn • Boston • Cincinnati • Detroit • London • Madrid
Melbourne • Mexico City • New York • Pacific Grove • Paris • San Francisco
Singapore • Tokyo • Toronto • Washington

NOTICE TO THE READER

Publisher does not warrant or guarantee any of the products described herein or perform any independent analysis in connection with any of the product information contained herein. Publisher does not assume, and expressly disclaims, any obligation to obtain and include information other than that provided to it by the manufacturer.

The reader is expressly warned to consider and adopt all safety precautions that might be indicated by the activities described herein and to avoid all potential hazards. By following the instructions contained herein, the reader willingly assumes all risks in connections with such instructions.

The publisher makes no representation or warranties of any kind, including but not limited to, the warranties of fitness for particular purpose or merchantability, nor are any such representations implied with respect to the material set forth herein, and the publisher takes no responsibility with respect to such material. The publisher shall not be liable for any special, consequential, or exemplary damages resulting, in whole or part, from the readers' use of, or reliance upon, this material.

Cover Design: Paul Roseneck

Delmar Staff
Publisher: Alar Elken
Acquisitions Editor: Mark Huth
Editorial Assistant: Dawn Daugerty
Marketing Manager: Mona Caron

Printed in the United States of America

For more information, contact:

Delmar Publishers
3 Columbia Circle, Box 15015
Albany, New York 12212-5015

International Thomson
Publishing Europe
Berkshire House 168-173
High Holborn
London, WC1V7AA
England

Thomas Nelson Australia
102 Dodds Street
South Melbourne, 3205
Victoria, Australia

Nelson Canada
1120 Birchmont Road
Scarborough, Ontario
Canada M1K 5G4

International Thomson Editores
Campos Eliseos 385, Piso 7
Col Polanco
11560 Mexico D F Mexico

International Thomson
Publishing GmbH
Königswinterer Strasse 418
53227 Bonn
Germany

International Thomson
Publishing Asia
221 Henderson Road #05-10
Henderson Building
Singapore 0315

International Thomson
Publishing–Japan
Hirakawacho Kyowa Building, 3F
2-2-1 Hirakawacho
Chiyoda-ku, 102 Tokyo
Japan

1 2 3 4 5 6 7 8 9 10 XXX 03 02 01 00 99 98

Library of Congress Cataloging-in-Publication Data
Bevelacqua, Armando S., 1956-
Terrorism handbook for operational responders / Armando Bevelacqua, Richard Stilp.
p. cm.
Includes index.
ISBN 0-7668-0475-5
1. Terrorism. 2. Emergency management. I. Stilp, Richard H.
II. Title
HV8431.B49 1998
658.4'73—dc21 98-7415
CIP

CONTENTS

QUICK REFERENCE INDEX

Introduction

Contemporary acts of terrorism gravely concern lawmakers responsible for public safety. Their concerns are real: hazardous materials and individuals willing to use them for terrorist goals increasingly proliferate in our society. Terrorism and the acts that surround "terrorism" have occurred within civilization for millennia. However, there have never been so many incidents from such a variety of groups worldwide, each operating within its own niche.

International terrorism has been a global issue and threat for decades. France, Ireland, Italy, Israel, Bosnia, Libya, Iran, and Iraq have been dealing with these "freedom" fighters for years, since political unrest and/or differences create groups that desire to change their environment. The United States also has a long tradition of freedom fighters. Look back at the Revolutionary War or post-Civil-War eras. Each fostered the growth of militia groups willing to fight for the validity of their causes. Some of these same groups still have members today. Domestic groups and international factions working within their respective cultures each have their own axe to grind. However, domestic terrorism has been redefined. Recent targets in the United States have included the World Trade Center in New York City (February 1993, linked to Middle Eastern factions), the Murrah Federal Building in Oklahoma City (April 1995, linked to the Waco, Texas, incident a year before), Anaheim, California (threat directed at Disneyland), the random bombing at the Summer Olympics in Atlanta, Georgia, in 1996, and a variety of militia and extremist activities around the country. The threat of a terrorist incident is credible.

The essence of terrorism motivated by political idealism—its unanticipated, premeditated, intimidation by force and total disregard for human life—disturbs individuals. As defined by the New World Liberation Front, a terrorist incident is "an action that the urban guerrilla must execute with the greatest coldbloodedness, calmness, and decision." Unfortunately, it is a crime that, with the violence of rape and the brutality of murder, erodes the infrastructure of modern society.

Response to acts of violence has primarily been a law enforcement issue. However, as the terrorist activity in the United States has shown, response from all emergency services is required. This has posed a unique challenge to the emergency responder. Such intentional acts of violence are not limited to nuclear, chemical, or biological weaponry. Common hazardous substances, hazardous waste and its illegal disposal, and detonation and use of warfare agents are all potential weapons for the motivated terrorist. They are all the same in terms of emergency response. Although each may be handled in a unique fashion, all have the potential to affect and erode the infrastructure of any community.

The affiliation of the terrorist sect, where it is from, or what its political agenda may be does not matter. Training for emergency response personnel must occur here and now on various levels, in many jurisdictions, and within operational areas of emergency response. Education of the first responders is the best defense against these unthinkable events. Thus, the purpose of this book is to give first response personnel operational direction during the aftermath of a terrorist incident.

The manual is organized into operational chapters; the chapter on establishing risk and response, a process that should occur before the event, is first. Chapters on nontraditional hazardous substances, their dangers, treatments, and the protocol used to handle them follow. Protocol is divided into basic treatment and advanced treatment, enabling the reader to choose the level of response suited to a particular situation. Following the hazard sections are chapters on the issues and problems that surround large-scale terrorist events. The appendix is a quick reference to the body of the text; separate reference cards correlate information presented earlier. The Glossary and Index cross-reference acronyms and terms that may be unfamiliar to you.

Emergency responders must wage a war for our own personal safety and welfare as well as the safety and welfare of those we are sworn to protect. Through knowledge, training, and planning, we will safely overcome the obstacles emergency work presents. This manual is dedicated to all emergency responders here, now, and in the future. The time to read, plan, and implement is now, before the event. During or after will be too late. Good luck, Godspeed and operate safely at all your emergency scenes.

1

ESTABLISHING RISK AND PREPARING RESPONSE

OVERVIEW

Now that we know the outcome of Timothy McVeigh and Terry Nichols' trials, all Americans can rest easily knowing that justice has prevailed. In fact, those of us involved in emergency response may give a a sigh of relief, feeling that we can now close the book on terrorism in our country. Wrong! McVeigh and Nichols have only completed the first chapter of what many believe will be a novel filled with terror, set during the next decade. In fact, law enforcement officials have stated that every major city in the United States will respond to terrorist events in the very near future. Why America and why now? These questions are difficult to answer, but the trend is definitely in this direction.

The United States government has responded by organizing and establishing strike teams to train local responders in the management of terrorist incidents. This ongoing process has required millions of dollars to train both specialized military units and civilian counterparts alike. You may think that all of this activity should only be focused on large urban areas. You may also believe that the attention given to this subject, because of a few isolated incidents, is far more than it truly deserves. If this is your belief, we sincerely hope you are correct and the preparations being made go unused.

However, let us take a few moments to increase your understanding of the magnitude of terrorism. On March 20, 1995, at 0809 hours, a terrorist attack took place in a Tokyo subway. A radical group, led by Shoko Asahara, the Aum Supreme Truth, deposited several vinyl bags containing a poisonous chemical, which produced deadly sarin (GB, a nerve agent) vapors on three subway lines. The chemical vapors, spread rapidly throughout the stations. The Tokyo fire department responded throughout the morning by dispatching a total of 340 units staffed with 1,364 fire rescue personnel to the scene. In total, they responded to 688 patients. More than 135 firefighters were overcome with the vapor and required medical care. By the end of the incident, there were 5,510 victims; twelve died (*Time* 1995:145:14—The reign of terror started in June of 1994, when sarin gas was released from a car traveling through a neighborhood in Japan. The neighborhood was home to several judges who were to make a legal decision on land owned by the cult.). This attack was extremely easy to execute and had devastating results. Can you consider your department prepared? Is your community ready to handle an incident that will totally tax the resources available? Do you have a plan of action? These questions and many others must be answered before any community can begin to consider itself prepared.

Some have identified terrorist incidents as hazardous materials events, others have described them as low-intensity military conflicts, while still others have represented them as multiple casualty incidents. No matter how they are classified, it is more than likely that they will tax any emergency response system far beyond its capabilities. Command and control are the keys to successful mitigation of these events. Preplanning should include an understanding of weapon types and potential solutions to problems presented by an attack. If a terrorist is to carry out an attack successfully, he/she must overcome challenges in the form of technological and political barriers.

BARRIERS

Technological Barriers

Technological barriers are obstacles that an individual must pass in order to use a nuclear, biological, or chemical agent as a weapon of mass destruction. Take, for example, a biological agent that, in theory, can be created in the terrorist's own garage. The technological barriers that must be overcome to get to the completed goal are many. The first barrier is acquisition of the pathogen, but for argument's sake, let us assume that acquiring a biological pathogen (first barrier) and maintaining that pathogen as a viable organism could be done without too much difficulty. The next hurdle would include the cultivation and maintenance of these pathogens in a sufficient quantity to deliver a weapon capable of affecting the target. This alone leads to several additional obstacles:

1. A containment area maintaining optimal growth temperatures and medium must be constructed. Workers in this biological lab must maintain high-level protection using quality personal protective equipment. Above all else, the laboratory must remain clandestine enough to escape discovery.
2. Within this laboratory, highly skilled workers would be required to cultivate and maintain the organisms in sufficient quantity to be used.
3. Once this is achieved, a dispersal system must be constructed, bought, or otherwise acquired for product delivery.
4. Once the above obstacles are overcome, the terrorist must successfully deliver the weapon to a target. In the case of biological weapons, personnel delivering the weapon may have to wear high-level personal protection, drawing additional attention to themselves in a crowded place. This, in itself, presents further obstacles.

Once released, the weapon could induce uncontrollable effects on a given population, resulting in an epidemic or pandemic spread. The results of an action such as this would be catastrophic and could include the death of the terrorist by either intentional or accidental means. The goal of a political terrorist is to destroy the public's faith in the government's ability to protect its citizens. If this goal is attained, the result will be distrust of the government and sympathy toward the terrorist's philosophy.

Choosing chemical agents as products of mass destruction poses similar obstacles that, again, must be overcome. Educational expertise must be available during the acquisition of product precursors, system design, and manufacturing. Manufacturing and refining the product require a high level of personnel protection, equipment, and physical space. Implementation of the product into use requires systems designed to disperse it over the intended target.

Nuclear devices create, for the terrorist, an additional set of complications prior to weapon development and deployment. The product itself, a radioactive isotope, is severely hazardous to the handler. These isotopes require an extensively protected storage vessel that, in itself, may weigh tons. Designing and building a thermonuclear explosive device takes extensive expertise, logistics, and almost unlimited resources. Only a few countries have succeeded in building one. This type of device development would be highly unlikely in terrorist organizations that usually operate under very limited budgets and even more limited expertise.

In reality, a terrorist nuclear threat would more likely be in the form of an explosive device used to disperse radioactive material over an area or population. This type of device is typically called a "dirty bomb." A dirty bomb would probably provide the level of destruction and population anxiety that a terrorist would want to achieve without the technological barriers associated with a thermonuclear device.

This leads to the last category of mass destruction weaponry—munitions and incendiary devices. In actuality, this one category represents 70 percent of all terrorist strikes. This type of weapon of mass destruction creates an immediate impact from easily acquired materials. It requires low technological expertise, can be effectively placed and detonated by one individual, and minimizes the risk to the user. Furthermore, the use of munitions attached to industrial processes, chemical storage containers, nuclear products, and even biological hazardous materials all represent a possible means of dispersing agents of a chemical or biological nature.

In many cases, the terrorist's resources, or lack thereof, is the only factor that limits their desired effects. Occasionally, the desired outcome may be intentionally limited to display a show of strength and not to create complete devastation. This effect can cause as much terror as an all-out attack using fewer resources.

Political Barriers

Political barriers are yet another obstacle that influences the type of terrorist activity attempted. International terrorists with political motives are considered to be patriots, heroes, or freedom fighters within their own country or political circle. In these cases, it may be the terrorists' own country that is sponsoring their attacks. At other times, the terrorists' acts of violence are not condoned by their own country and they become criminals in their own homeland.

The recent fall of the Soviet government made the United States the most powerful military force in the world. This political change has lead to an increased threat of terrorism in America. Small countries in opposition to U.S. ideals would historically look to the Soviet Union for military support. Now that these small countries have no superpower to back their political stance, and their military is no match for the U.S. military, terrorism may be the only means they have to flex their muscles. Threats of terrorism from other countries and foreign militant groups have been common. Few have, to this date, been carried out.

In general, three types of "patriots" can be identified:

Type 1 includes those individuals who function as groups and work across the globe as military entities with a full range of military tactics. These are sometimes called commandos or professional mercenaries ("Mercs").

Type 2 includes those groups that have limited military capabilities, and, usually, limited military tactical knowledge, but that are able to function as pseudomilitary combatants. These are such indigenous groups as drug lords, militias, and psychopathic cult groups, which believe that political change can be accomplished through fear generated during low-intensity conflicts.

Type 3 includes individuals not affiliated with recognized groups, who serve as amateur patriots in the name of a given ideal. These individuals can work in groups or as unattached political subversives.

The profile of a terrorist is found in any one of these categories or combination of categories. A true terrorist predicts and calculates how much violence to use in making his/her statement. If a terrorist act creates too much anguish in the targeted society, support for their goal is eliminated, and the act may result in retribution and extreme punishment. But if the terrorist attack is carefully calculated, the desired effect results in more government restrictions on the population, a perception of the government controlling the lives of its citizens, and civil unrest. This is what the terrorist works to achieve.

Political terrorists are motivated by strong beliefs and use violence to persuade others to believe the same. Violence may come in the form of bombings, poisonous gas release, or the use of biological weapons. Terrorists may accomplish their ends by devising their own weapons and sabotaging industrial settings or transportation vehicles to obtain the desired effect.

Such books as *The Poisoner's Handbook*, *Silent Death*, *Silent Tool of Justice*, *Uncle Fester*, *Assorted Nasties*, and *Get Even: The Complete Book of Dirty Tricks* have cookbook recipes necessary to brew deadly poisons or devise homemade bombs. Other toxic materials can easily be made with minimum knowledge and effort. Ricin, for example, can be made from castor beans, is 6000 times more toxic than cyanide, and has no antidote. Furthermore, it is hard to detect during autopsy, because it is a simple protein difficult to isolate. Ricin, in quantities large enough to wipe out a small city, has already been confiscated from militia groups in the United States. With Internet access, any curious villain can locate enough information to cause destruction in our communities or gathering places.

Bombs have long been the weapon of choice for terrorists. From 1990–1995, 15,760 bombings have occurred. In total, 355 individuals have been killed, with 3,176 injuries. The monetary damages from these criminal events has been in excess of 600 million dollars. As you can see from these statistics, terrorism has affected a small number of victims with high monetary impact. It is not always the loss of life or destruction that gives the terrorists their edge, as natural illnesses and accidents cause more death and destruction. Instead, the indiscriminate nature of the attack causes a psychological impact that destroys the public's feeling of natural freedom. This is the desired goal that drives the terrorists' attack.

COMMAND AND CONTROL

Command and control of terrorist events must be handled much differently than other emergencies. A militaristic attitude on the part of the first responders and commanders must be maintained while fire and rescue objectives are accomplished. When bombings are the weapon of choice, the commander must be concerned about secondary devices, structure stability, search and rescue capabilities, and the possibility of another type of strike while rescue efforts are underway.

During a chemical or biological attack, other issues must be addressed. These issues include concerns about the type of chemical or biological agent used, level of protection necessary for successful mitigation, detection capabilities, and decontamination procedures. In both cases, the level of critical infrastructure destruction and the continuity of local and state government response all necessitate more sophisticated command and control structure. Legal implications, law enforcement necessities, and evidence acquisition are additional stressors placed on the shoulders of the incident commander. It is for these reasons that a unified command structure is necessary for controlling an incident of this type. The responsibilities are not purely fire/rescue and EMS but also involve law enforcement. Remember that terrorism is a crime that causes injury or death to innocent persons.

Initial command information should include the condition of damaged structures (and exposures), evacuation needs, number and location of casualties, and how long the response times will be for technical assistance. Initial concerns should also include an understanding of the capabilities of all emergency responders, technical rescue teams, mutual aid agencies, and their resources. Objectives should always include strong security measures and evidence preservation.

During any terrorist event, harsh decisions may have to be made that go against traditional standard operating procedures. A chemical or biological terrorist attack may necessitate a decision to exclude rescue workers from entrance into the "hot zone," which becomes an area of acceptable casualities. This is a military concept (termed "kill zone") used to identify the area of contamination. Emergency responders do not consider acceptable casualties as a viable measure, and are not likely to use it unless trained and prepared for such extremes. Unfortunately, this may lead to additional life loss within the emergency response community.

Evidence conservation is an important issue and may involve everything from the identification of debris lying long distances from the incident to acquisition of a victim's clothing. Emergency responders must be taught to preserve evidence, taking care not to destroy or discard anything. Even clothing cut from contaminated and injured victims must be bagged, marked, and maintained.

Expedient responders (those citizens responding to help, for example, doctors or nurses) wishing to assist must be coordinated through the command center. In recent cases, too many hospital employees responded. This left hospitals shorthanded and in need of assistance. Unless requested to respond, expedient responders should stay away so their help can be better coordinated. Remember that the only rescuer death at the Oklahoma City bombing was a nurse working as an expedient responder.

PREDICTABILITY

A terrorist attack is an extremely unpredictable act that can occur anywhere, usually without warning. The World Trade Center and Oklahoma City bombings occurred without warning, and were both intended to destroy physical property and human life. Preplanning, an essential function of the fire service, is of some benefit in the preparation for a terrorist attack, but because of the unpredictability of these events, communitywide preparation is more beneficial. Preplanning for terrorism must include the identification of critical infrastructure and services such as telecommunications; electrical power systems; fuel/oil production and storage; transportation corridors; processing facilities; banking and financial organizations; water supply systems; and emergency services. All of these constitute areas of possible attack. Once identified, emergency response agencies and local governments can determine risk assessments for their communities and formulate an action plan should an attack occur.

Location, occupancy, timing, and type of gathering all contribute to the possibility of a terrorist incident (see Appendix A). Once an incident occurs, emergency responders must proceed with caution and have a high degree of suspicion. Secondary devices have been encountered and are placed to hamper and harm emergency responders. Incident commanders must treat any incident as an active bomb scene, hazardous materials incident, multiple casualty incident, and military campaign. Proceeding with such great caution may be foreign to many emergency response personnel who are anxious to get to work once an incident is identified. However, it is important to remember that this is not a typical rescue response, but, instead, a low-intensity conflict that may include additional antipersonnel weapons aimed at those attempting to rescue the initial victims.

To avoid predictability, staging locations for target buildings should be rotated to avoid setting a pattern that may be easily seen by a terrorist. Even varying response routes is prudent, especially if multiple threats have occurred or intelligence from the police indicates that a higher degree of threat is possible. Our predictability may place us at greater risk.

MEDIA CONCERNS

As with any major incident, good media interaction with command is necessary for accurate, appropriate information to be released. Remember that many terrorists attack to bring attention to their cause; in this respect, the news agencies contribute to the completion of their objective. The media can also control the number of expedient responders on the scene. Getting information to concerned individuals about the dangers of responding to an uncontrolled scene, and notifying off-duty hospital personnel that their services would be more useful in the hospital or in remote medical units are positive media contributions.

CURRENT TRENDS

Current trends in training and organizing both federal and local emergency responders have been stimulated by the enactment of Presidential Decision Directive-39 (PDD-39); the Nunn-Lugar Act. Crisis management is the responsibility of the

Federal Bureau of Investigation (FBI), while consequence management is the responsibility of the Federal Emergency Management Agency working closely with the FBI. This designates primary responsibility for emergency response to the federal response plan administered under the various Emergency Support Functions (ESF). If an incident occurs, resources available to local and state governments are acquired by request for federal disaster declaration. This request is accomplished when the local government contacts state government, which in turn notifies federal agencies who activate the Federal Response Plan (FRP).

Metropolitan Medical Strike Teams (MMST) are being developed, funded, and trained by the federal government to assist local response agencies immediately. These teams, like their counterparts (Rapid Assessment Intervention and Detection [RAID], Disaster Medical Assistance Teams [DMAT], Urban Search and Rescue [USAR]), are geared to support specific activity during a terrorist event. Because of their level of training in personal protective equipment, decontamination procedures, and the handling of mass casualties, hazardous materials teams are being identified as the response individuals for such a job. Regional hazardous materials teams have also been developed to handle routine chemical emergencies and to assist with MMST functions. DMAT and USAR also have roles within the terrorist event response; both are identified as resources for incidents of mass destruction.

■ SUMMARY

Terrorism using explosive, chemical, or biological weapons is not a new problem. History reveals many instances of the use of these weapons. The question that must be asked in present day is how ready are our emergency services and public health systems to deal with an incident involving these weapons. The first line of defense against terrorism is intelligence gained by law enforcement agencies. This intelligence is what will forewarn emergency responders that an unconventional incident may occur. Other clues may also play a part in forewarning emergency responders. For example, instances of foreign political unrest involving unpopular American intervention may stimulate terrorist attacks in the United States.

A report on threat conditions (TCON) is normally acquired by high-level law enforcement agencies capable of providing information to local emergency responders. Historically, the information gained by law enforcement agencies has not reached fire and emergency medical responders. So for fire and emergency medical services, intelligence about an increased threat of attack is low. In the same community, police agencies are geared for a possible incident. Communication channels must be opened between law enforcement and fire/emergency service agencies to provide important intelligence to all who have a stake in emergency control and mitigation.

At best, control of any incident is challenging. Through the use of incident management, command can control working forces in a coordinated, efficient manner. At a terrorist incident, successful command of these forces will depend on training level and preplanning completed prior to the event. Since a terrorist event involves police, fire, and emergency medical services, an integrated/unified command system must be initiated from the onset. Control also depends on quickly identifying hazards and instituting appropriate solutions to deal with them.

With the potential use of weapons of mass destruction, including high-power munitions and chemical or biological agents, multicasualities and extreme loss of life are possible. Historically, terrorist events have not produced high-level kill ratios in the United States. Even internationally, most historical terrorist events have not involved a tremendous loss of life, and have purposely excluded women and children. Events such as the Lockerbie, Scotland, airline crash, where hundreds of innocent victims were killed as the result of a bombing, are rare occurrences. The intent seems to have been to strike areas that would trigger media attention and cause fear in selected populations, but not to kill multitudes of innocent persons. In recent years, however, this trend seems to be changing. The World Trade Center and Oklahoma City federal building explosions were attempts to kill large numbers of people. The sarin attack in the Tokyo subway had the potential to kill thousands, but was, for the most part, unsuccessful. The rules in recent attacks have changed. This new trend does include striking with intent to cause a large loss of life regardless of age or sex.

As in any disaster, the speed and effectiveness of responders and efficiency of command are the keys to limiting the loss of lives and critical property. Because terrorist activities are often directed toward the loss of human life, the initial focus of response must be on the health and medical consequences of the incident.

For successful implementation of a diversified emergency response plan, certain activities must be initiated and maintained within all emergency services:

1. Develop integrated training and exercise activities that include fire, police, EMS, hospital, health departments, emergency management organizations, and local/state agencies.
2. Promote integrated federal, state, and local planning toward regional response teams that include MMST, RAID, DMAT, USAR, FBI, Metropolitan Bureau of Investigation (MBI), and local police department and fire department special operations.
3. Ensure sufficient medical supplies, personal protection equipment, decontamination equipment, evidence recovery equipment, and stress debriefing to responders and hospitals for use during and after the incident.
4. Develop integrated informational intelligence systems between police and fire departments, including existing agencies such as the MBI, FBI, Centers for Disease Control (CDC), health departments and health entities (both state and local), and public health advisory councils as repositories of information during such a crisis.

Terrorism is becoming a major problem in this country. It is as obvious as two tons of explosive parked in front of the Oklahoma City federal building or as silent as a container of ricin in the hand of a would-be assassin. It is as real as a pipe bomb exploding at an abortion clinic and as grim as recruitment videos for militia and racist supremacy groups. Unfortunately, the tools of terror have become less expensive, more accessible, and incredibly more lethal than ever before. Time has come for our preparation. The list of needs and necessities are long and complex. Issues involving response across agency borders must be set aside and agreements made for mutual response. Understanding the complex interdependency between all agencies is an essential part of the response equation. Working together for a systematic multilayered response system, local, state, and federal agencies can make great strides toward addressing deficiencies in emergency response capabilities. To do anything less would be a disservice to the community. As an Irish Republican Army terrorist once said, "I have to be lucky only once; you must be lucky every time." We must attempt to be smarter and better prepared so that we can be lucky all of the time!

2

MUNITIONS

OVERVIEW

As with any emergency, scene size-up starts from the time dispatch assigns the call and does not end until departure from the incident. It is a continuous assessment of the incident and all of its components. Incidents involving terrorist acts are no different than other emergencies. Just as in fire or hazardous materials emergencies, there are clues as to the type of incident and control measures necessary to manage the incident.

Location and occupancy should give the emergency responder clues to the appropriate action profile. Schools, public assemblies, transportation facilities, and politically sensitive businesses are all favorite targets for bomb threats and bomb detonation. When suspicious calls to these signature targets are received, they should alert responders to the possibility of a terrorist attack. These particular targets stimulate excessive media attention that feeds the goal of terrorism—FEAR. In today's political climate, company officers, dispatch centers, and city management should be highly suspicious when threats are made toward these targets.

According to the FBI, the definition of a terrorist event is "the unlawful use of force against persons or property to intimidate or coerce a government, the civilian population, or any segment thereof, in the furtherance of political or social objectives." (PDD—39; FBI). The choice of how terrorists use force is solely dependent on the affordability of the weapon, their ability to move and disperse the weapon, their level of technology (usually low), and their ability to deny the results if the intended objective was not achieved. With this in mind, it is easy to see why bombs have been the weapon of choice for terrorists. Bombings enable terrorists to make political statements through the media and cause a fear response associated with a specific target. In short, bombs work very well to fulfill terrorist goals. They are inexpensive, simple to build, easy to place, and can be assembled from commonly obtained materials. Bombings become high-profile media events with sudden impact, resulting in destruction of political confidence.

BOMB INCIDENTS

Bombs and incendiary devices cause a majority of terrorist incidents. Bombings are responsible for approximately 70% of the historical occurrences of terrorist activity. They are not the only tools used by these criminals; they are just used more often.

The Department of Transportation (DOT), for the purposes of transportation, defines an explosive as any substance or article, including a device, which is designated to function by rapidly releasing gas and heat. DOT further states that an explosive may include a chemical reaction that rapidly releases gas and heat, unless the substance or article is otherwise classified. These definitions are used to define placarding, labeling, and capability standards for transport. Under most conditions, the law-abiding transporter will comply with the transportation criteria defined by DOT. Obviously, terrorists are not concerned with law and will go to great lengths not to bring attention to what they may be transporting. Therefore, responders may encounter any type of device or material being illegally transported by those intending harm. Improvised munitions mixed or assembled haphazardly add a much higher degree of hazard to an already dangerous response. Timing devices, poor quality precursor chemicals, and unpredictable chemical reactions are examples of extreme hazards to expect from improvised munitions.

Incidents involving bombs may have three stages. These stages are:

1. The threat of a device is received.
2. A device or suspicious package is located.
3. An explosion and its aftermath occur.

Although it is worthwhile to review each stage of an event of this nature, it is also important to understand that every terrorist event is different and unpredictable. The terrorist may choose not to issue a threat, but, instead, the surprise nature of the attack may add to the fear stimulated after the bombing has taken place.

■ BOMB THREAT

The most difficult incident to deal with is the threat of a bombing. Notification may come in many forms: mail, a telephoned message to the location, media notification, or handwritten messages on walls or bathroom mirrors. Handwritten messages such as these were done in the 1970s during the hijacking of aircraft. During these hijackings, the messages provoked the fear of bombs aboard the aircraft. They also raised the activity level of emergency response personnel. Activation of emergency response personnel further raised the level of fear and added to the potential effects of the threat.

It is difficult to assess the risk of a bomb threat. Overreacting can prove to be expensive and disruptive to business, and may play right into the hands of the terrorist. Underreacting can place more lives in danger and, thus, be even more costly. Bomb threats are the most successful means of terrorism, because the desired result of disruption and fear is immediately evident (at least to the area of society the terrorist is trying to affect). Once an explosion has occurred and disruption has been achieved, media sensationalism then furthers the terrorists' goal of drawing attention to the event. Terrorists may believe that a warning call, followed by an explosion that kills victims, is proof that the government cannot protect its citizens. In the eyes of some citizens, this failure to protect their safety and well-being is a betrayal of the government's promise to keep them safe.

■ DEVICE IN POSSESSION OR SUSPICIOUS PACKAGE

The significance of an actual or suspicious device is clearly recognized by emergency response personnel. However, due to the increasing frequency of acts of terrorism, emergency response personnel have become more sensitized to the reality of such a device. Added to the challenges facing emergency responders is the technology now available for constructing these devices. For example, complex electronic timing devices have been used to trigger a device and have even been programmed to activate the explosive prior to the reported detonation time. This serves to fool emergency responders, lulling them into believing that minutes or hours are available, when, in reality, they are in immediate danger. The smells of gunpowder or signs of smoke (usually brownish-orange) may signify that detonation will occur shortly. During this scenario, it is too late to evacuate the building; self-preservation should be the only goal.

As part of the sizing-up process, the target profile should be considered. History reveals that there is a higher frequency of bombings in July, August, and September. It has also been determined that they occur most often on Monday. Knowledge of these statistics and evaluation of the target profile can give responders clues as to the probability of a bombing event. The target profile should take into account three basic elements:

1. location and occupancy,
2. primary event type, and
3. timing of the incident.

Once emergency units arrive, continuous scene size-up should progress into scene management appropriate for a terrorist event (see Appendix A).

Responders must become familiar with the devices, agents, and overall tactics that are employed by terrorists. During several recent events, the tactic of placing secondary devices (or antipersonnel munitions) to injure emergency responders was used. These secondary devices are placed in areas known to be staging or parking locations for emergency vehicles. In these cases, the primary device caused damage and attention at the scene, stimulating an emergency response. Once an audience of curious onlookers and emergency personnel assembled, a second device was detonated. These have been termed by emergency responders as "sucker punches," because they are intended to surprise and hurt or kill responders. Terrorists known to have practiced this tactic study emergency response staging areas to get the best effect from their sucker punch.

So how do emergency responders protect themselves against this type of devious act? They must make a conscious effort to become aware of possible hazards in their communities, watch for trends across the United States, and approach suspicious incidents with the knowledge that hidden hazards may exist. Some simple practices that can lessen the chances of becoming injured on an incident include staging apparatus and staying clear of

- large shrubbery groupings,
- commercial garbage dumpsters,
- mailboxes, and
- other areas that could conceal a secondary or tertiary device.

Another concern about operating at a scene where either a threat of a bombing has occurred or there has already been a detonation involves the use of radio wave transmitting devices. Any communication in the area of a suspected device, including hand-held radios, cellular phones, and mobile data terminals, is strictly forbidden. Radio transmissions and static electricity have sufficient energy to produce electrical current capable of initiating ignition. Guidelines for operating radio transmitters at a suspected bomb incident are as follows:

- transmitter wattage of between 5 and 25 W should have a stand-off distance of 100 feet;
- transmitter wattage of between 50 and 100 W should have a stand-off distance of 220 feet; and
- transmitter wattage of between 250 and 100 W should have a stand-off distance of 450 feet.

Generally, good practice should include a 1,500-foot minimum isolation distance for radio equipment. Many jurisdictions have identified a radio isolation distance of 2,000 feet as a standard operating procedure. This radio-free area must be maintained in order to ensure responder safety.

Establishing a communication link between the hot and cold zones will provide a formidable challenge to emergency response personnel. Additionally, this communication link will require the support of other responders, for example the placement of a spotting scope at both ends in order to communicate. This may in fact involve a variety of hand signals or Morse-code-type communication.

Immediate evacuation and isolation from a real or suspected bombing incident should be as far as 1,500 feet. When dealing with military ordinance, the suggested evacuation and isolation distance can be as far as 5–10 miles. Dispatch and/or incident command should transmit the order to turn off radios once the units or personnel reach the staging area. Radio-traffic-free areas, like staging areas, should be identified. If this order is not given, all transmissions in the hazard area (hot zone/warm) should be prohibited.

The type of munitions and the size of the package, if known, should play a part in action plans in terms of evacuation distances, resource allocation, and expertise needed to mitigate the incident. Additional factors may play into the action plan, such as the training level of responders and extrinsic hazards like booby traps. These may be in the form of a smaller explosive or other types of antipersonnel devices. All devices must be taken seriously, no matter how large or small. Formulas to build explosive devices are easily acquired, and the materials are inexpensive and commonly found. For example, the bomb used at the Oklahoma City federal building was primarily assembled with fertilizer-grade ammonium nitrate. An example of mixing toxins in explosive material occurred during the Iraq conflict, where reports of munitions containing aflatoxin were used against American troops. (Aflatoxin causes liver cancer months to years after exposure.) The possibility of toxic or infectious materials used in conjunction with explosives presents hazards with endless possibilities.

■ ANATOMY OF EXPLOSIVES

Improvised explosive devices may contain filler material rated as high- or low-filler explosive material. High-filler materials are further described as high-order or low-order detonation substances, the difference between these materials being the degree of consumption during the explosion. High-order materials detonate using all of the filler material at a burning rate of 3,300 feet per second (speed of sound), destroying the target by shattering structural materials. Low-order materials deflagrate (rapid burn) at a rate less than 3,300 feet per second (1,100 fps is the overpressure blast wave) and may not completely consume the filler, resulting in incomplete detonation. Low-order fillers destroy targets by a push-pull-shove effect, weakening structural integrity. Low-order detonations are hazardous for the first responder, because if not fully consumed, they leave explosive and even shock-sensitive material scattered around the scene. The pressure waves for high-order detonation reach 50,000–4,000,000 psi, while low-order detonation pressure waves reach up to 50,000 psi.

High-order filler explosives follow a sequence of events to detonate. This sequenced set of events is called an explosive train and has three separate segments: ignition source, initiating explosive, and main charge. Between the initiating explosive and main charge, a booster charge or delay charge may be used to ensure complete consumption. See examples of explosives on page 16.

The rate of burning has a relationship to the shrapnel propelled outward. Roughly, the rate at which shrapnel is pushed away from the explosive device is approximately 75% of the rate of burn. For example, if a substance burns at 19,000 feet per second, then the shrapnel will move at 14,000 feet per second. If the charge has a large piece of metal, the direction will be outward away from the detonation, causing further damage.

Common initiating explosives include either electric or nonelectric blasting caps to start the train. Electric blasting caps consist of a shell with several explosive powder charges and an electric ignition element. Electrical caps are dangerous because of the possibility of accidental ignition from static electricity or energy from radio transmissions. (Radio transmissions have enough potential energy to start the train.) Nonelectrical blasting caps use a fuse and initiator for ignition. A safety fuse, for example, has a combustible primer charge at the insertion end.

■ EXPECTED EFFECTS FROM EXPLOSIONS ■

Potential Injury	*Pressure psi*	*Structural Effects*
Loss of balance/ rupture of eardrums	0.5–3 psi	Glass shatters; façade failure
Internal organ damage	5–6 psi	Cinderblock shatters; steel structures fail; containers collapse; utility poles fail
Pressure causes multisystem trauma	15 psi	Structural failure of typical construction
Lung collapse	30 psi	Reinforced construction failure
Fatal injuries	100 psi	Structural failure

■ EXPLOSION AND AFTERMATH (PHYSIOLOGY OF BLAST EFFECTS)

When a detonation occurs, a wave of pressure moves from the detonation point outward in all directions. The initial shock wave is dependent on the type of explosive, confinement of material, and the oxidizers present. This pressure continues in every direction until the released energy has been equalized. Depending on the level of explosive order, a pressure wave can reach well above 50,000 psi. Think of this wave as a locomotive hauling freight that strikes objects while moving at 15,000 miles per hour! Atomization and total disintegration of material, including all biological material, will occur close to the detonation.

The primary pressure wave produces a shearing effect in such organs as the gastrointestinal tract, eardrum and surrounding bones, lungs, and central nervous system. The secondary wave causes material to fly around, striking victims, and causing them traumatic injury. Blunt trauma, lacerations, abrasions, puncture wounds, penetrating injuries, and incisions are all common. The tertiary effects of the blast are due to deceleration injuries. Once the explosion occurs, the victim is thrown in the direction of the initial blast wave.

■ EXAMPLES OF EXPLOSIVES ■

High-Order	
Trinitrotoluene (TNT)	Pale yellow solid, light cream to rust in color
Tetrytol	Alternate to TNT, 2, 4, 6-trinitrophenylmethylnitramine, very sensitive
Composition 3	Plastic explosive with an oily texture and yellow tint
Composition 4	A white pliable plastic explosive
Composition B	Mixture of cyclonite and TNT
Pentaerythritetranitrate (PETN)	A white solid, extremely sensitive, used in a detonating cord, equal in force to RDX and TNT
Amatol	Mixture of ammonium nitrate and TNT
Cyclonite (RDX)	A white solid, extremely sensitive, commonly referred to as RDX
Pentolite	Combination of Pentaerythritetranitrate (PETN) and TNT
Ednatol	Mixture of halite and TNT
Semtex	Pinkish color plastic explosive that can be molded
Picric acid	More sensitive than TNT, in the salt form bright yellow crystals
Nitroglycerin	Heavy, oily liquid that resembles water, generally pale yellow and viscous
Dynamite	

Low-Order	
Black powder	Mixture of charcoal, sulfur, and either potassium nitrate or sodium nitrate
Ammonium nitrate	
Pyrotechnics	Variety of fire works or chemical types, match heads
Incendiary	
Gasoline	
Pyrotechnics	
Butane, propane	
Hypergolic chemicals	Alkenes, alkynes in contact with a strong oxidizer

Blast Effects

The effects of a sudden release of energy from an explosion can be devastating. If a victim is far enough away from the center, so that disintegration does not occur, one of the following four general categories of injuries will be seen:

1. The first category of injury is caused by the initial blast wave, and they involve the hollow organs and their interfaces. The gas-filled organs are compressed and develop a gas-pressure exchange causing implosion injury. These injuries have a high mortality rate. See "Expected Effects from Explosions" on page 15.
2. The second type of injury is spalling. It occurs when pressure goes through a high-density organ then strikes a low-density organ, resulting in tissue violently spalling (tearing away from) off low-density tissue. These injuries have a high mortality rate.
3. The third type of injury involves organs that are attached or are in close proximity to one another. Target organs from this blast are abdominal viscera, lungs, nervous system, and skin. Moderate to high mortality rates result and survivability depends on the distance from the center of the blast.
4. The fourth category of injury is due to the toxic gases, dust, and fire created by the explosion. Similar to radiation emergencies, distance and shielding provide the best protection.

Management should be geared toward treatment of multiple systems trauma and deceleration injuries. If patients are contaminated, emergency medical treatment will be very difficult until decontamination is completed. If nerve agents or biologicals are used, technical decontamination may be nearly impossible (see Chapter 8, "Decontamination"). An explosion that occurs inside a sealed structure creates more trauma to victims compared to outside explosions. Outside explosions can disperse the energy easier and equalize the pressure more rapidly.

Multisystems trauma should be anticipated and treatment modalities should consist of volume replacement and shock management. Shock should be managed with high-flow oxygen, fluid replacement, positioning, and rapid transport. MAST

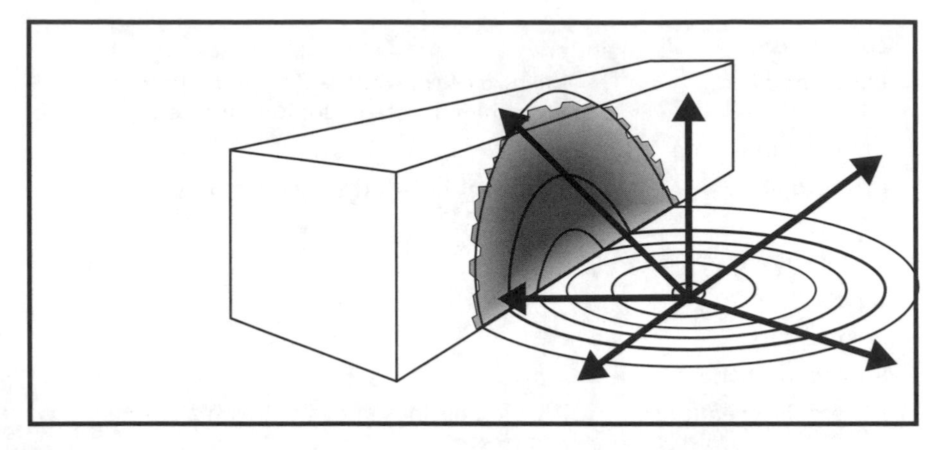

Figure 2.1 Direction of Shock Wave Travel

(medical antishock trousers) suits are contraindicated, even for splinting purposes, due to the possibility of crush injuries and decreased circulation. Normal post-explosion evaluations may include a twenty-four-hour hospitalization, even if there is no evidence of internal injury. Even a simple earache or headache may indicate deeper injuries.

Crush Injuries

Crush injuries and compartment syndrome are other injuries common to explosion events. Compartment syndrome is secondary to crush injury and is a condition in which the circulation of tissues within a closed fascial space (a thin membrane surrounding muscle groups) is compressed. Crush injuries are the result of a compressive force crushing the fascia-encapsulated muscle groups. They are caused by increased pressure exerted on an extremity from falling or flying objects. Compartment syndrome should be evaluated, for anytime there is evidence of a mechanism of injury or an obvious crush entrapment exists. Trapped victims located after an explosion should be evaluated for compartment syndrome related to the crush injury.

BASIC TREATMENT for crush injuries is primarily supportive. Recognition of compartment syndrome will be the greatest challenge. High-flow oxygen, notification of Advanced Life Support if available, positioning for shock, and rapid extrication are indicated. MAST suit application is contraindicated for crush injuries, even for splinting purposes, because it increases the pressure within the fascia compartments, furthering the syndrome.

ADVANCED TREATMENT

Pathophysiology consists of two events that occur either simultaneously or concurrent to one another. First is decreased size of the extremity due to external pressure. The second event occurs as a result of swelling that increases the compartment pressure. Once the compartment pressure overcomes the pulse pressure, circulation distal to the site stops. Crushed extremities should be continually evaluated for the onset of symptoms. Systemic complications occur rapidly and may consist of metabolic disturbance, renal failure, tissue necrosis, and infection. If lengthy extrication is necessary, the emergency medical provider should evaluate the patient for the onset of systemic symptoms.

As the syndrome progresses, muscle cell death occurs (rhabdomyolysis), due to increased pressure in the muscle fascia compartment. When these muscles start to die, a release of phosphorous, myoglobin, acid, and potassium occurs.

The release of potassium can be identified through the use of serial electrocardiograms (EKGs) before and during the extrication process. Enlarged "T" waves, prolonged QRS, differences in PR interval, P wave loss, and ventricular dysrhythmias may all result from increased blood potassium levels. Secondary to the increased potassium levels is the decrease in calcium levels within the cell. This in itself may cause cardiac arrest. Kidney damage and failure also result from the injury. The kidneys attempt to filter the myoglobin and acid from the bloodstream, which increases the acidity of the urine in the kidneys, precipitating the myoglobin and damaging the kidneys.

Treatment is limited to restoring local blood flow to the affected area. This is accomplished by eliminating any external pressure splints, limiting fascia pressure (fasciotomy or rescue amputations), using IV therapy to prevent renal failure, and using urine alkalization. IV (intravenous) solutions should be warmed to between 98–99 degrees Fahrenheit to decrease the occurrence of traumatic hypothermia.

Along with the previous protocols, surgical intervention is the only definitive method of treatment. When muscles are injured, swelling or edema occurs, resulting in ischemic tissue within four to eight hours. This is a direct result of increasing pressure. Once pressures of 25–40 mm Hg occur, rhabdomyolysis will begin. Prevention is geared toward reducing pressure and limiting metabolic toxins. Pressure can be reduced by performing a fasciotomy on the affected extremity or, in some cases, a rescue amputation. Consider fasciotomy as a prophylactic procedure if the mechanism of injury is associated with prolonged extrication time.

Vasodilator medication should be avoided, due to the fact that vasodilation has probably already occurred locally. Hypovolemia can be managed by large volumes of 9 percent sodium chloride solutions. Lactated ringers can be used but should be avoided if normal saline solutions are available. Initial management should start with a bolus of 1,000–2,000 cc with a maintenance drip of 300–500 cc per hour. This should be employed if the victim was trapped by debris and should maintain the patient until freed. Delaying fluid therapy *will* cause renal failure and death due to the precipitation of myoglobin.

Acidosis can be treated through the use of a sodium bicarbonate IVP adjusted to maintain a urine alkalization pH of greater than 6.5. Hyperkalemia should only be done after urine alkalization. Calcium chloride, 5–10 cc, followed by 25 cc $D_{50}W$ IVP is initiated by ten units of regular insulin IVP. Maintenance of diuresis at 300 cc per hour is recommended and managed by a Foley catheter. If a fasciotomy or rescue amputation is employed, ketamine can be used as an analgesic. Ketamine is given intramuscularly (IM) or intravenously (IV). As IM 4–6 mg/kg or IV slow 2 mg/kg, contraindications include intracranial pressure, age less than three, congestive heart failure (CHF), or hypertension. Consult your medical director for field use. A broad-spectrum antibiotic such as cephalosporin should be considered and will be indicated if rescue amputation or fasciotomy are employed. MAST (medical antishock trousers) suit application is contraindicated for crush injuries, even for splinting purposes, because it increases the pressure within the fascia compartments, furthering the syndrome.

■ SUMMARY

The aftermath of an explosion, regardless of size, will have a severe impact on local emergency services. Severe structural failure of the building of impact and surrounding exposures all may require specialized search and rescue teams such as USAR. Because of the overwhelming medical consequences, DMAT and MMST may be required to set up field hospitals, lessening the impact on local medical resources.

Dust created by the explosion may produce widespread exposure to chem-bio agents if they were employed in the device. After an explosion, normal dust production alone can cause widespread "gray" areas, making nonambulatory patients difficult to locate due to the camouflage effect. Fine particles of dust as small as one micron in size can penetrate the alveoli causing immediate injury. Penetration into the fine bronchioles by these small particles can also cause lung damage years later. Soot, a common by-product of a deflagration type of blast, can coat the lungs of a trapped victim, leading to an asphyxiating injury similar to those found in smoke-inhalation victims. During a large-scale explosion, dust particles suddenly blanketing populated areas can stimulate respiratory injuries, which, in themselves, can overtax emergency response systems and hospital emergency departments.

Law enforcement can also become quickly overwhelmed by controling the security of a large scene, and collecting debris, clothing, and fragments for evidence. Every attempt should be made by fire and emergency medical system (EMS) personnel to preserve evidence for law enforcement.

Mutual aid and outside resources can be initiated by contacting local emergency operation centers (EOC) and state warning points. Federal aid and assistance is only granted when appropriate notification has been made.

Early identification of resources needs to be planned and tailored for a variety of potential events. Disaster drills that consider potential needs and analyze resource capabilities and systems deficiencies at all levels of local government are critical to securing a positive end to a terrorist event. The time to plan is before the event. Once the incident occurs, plans no longer affect the outcome. Failure to plan and train begets failure and injury at the time of the terrorist strike.

3

CHEMICAL AGENTS

OVERVIEW

Most civilized people would consider that the use of chemicals to harm or kill innocent persons is inhumane, unjust, barbaric, and cruel. Nonetheless, world military organizations have used chemical agents since before World War I. Since that time, some agents have remained exactly as they were, others have gone through a metamorphosis to ensure their effectiveness. Because terrorists learn much from the military, responders should understand agents used by military forces in order to gain insight into these unconventional weapons.

These chemicals are classified in military terms describing their effect on the enemy. It is clear that the intention of the agents is twofold: to incapacitate and to kill. In this handbook, chemicals are identified according to their effect on bodily systems. When the intention of the chemical exposure is to cause death, a notation will be made in the text. This information is based on the properties of the chemicals and their toxic ratings.

For the most part, chemical agents are broken down into neurotoxins (military nerve agents), chemical asphyxiants (military blood agents), respiratory irritants (military choking agents), skin irritants (military blister agents), and antipersonnel agents (riot control agents). Because terrorists may use many other chemicals to complete their tasks, responders should understand that other chemicals may be used as weapons. Among these are anhydrous ammonia, hydrogen fluoride, and sulfur dioxide. All responders must be aware of the typical military chemical agents and the more common toxic gases and poisons used in industry.

NEUROTOXINS (NERVE AGENTS)

Neurotoxins are probably the most common agents selected for use in wartime activities. These agents are very effective because they can enter through virtually any route and cause severe incapacitation and death. They have been formulated to be extremely toxic to the intended target, but to break down rapidly so that invading troops can inhabit the area within days after an attack. Similar compounds used in civilian society are organophosphate pesticides; some of these possess extremely toxic qualities. Terrorists may choose to use commonly found pesticides rather than military agents to injure or kill intended targets. Parathion and tetraethylpyrophosphate

(TEPP), both commercial-grade insecticides, were used as military nerve agents prior to their use as industrial-strength insect killers. These agents have limited availability, but they are easier to obtain than military chemicals. Whether military or civilian chemicals are the weapons of choice, the physiology of poisoning is the same. The military are able to transport and use many of these chemical weapons safely by synthesizing two less toxic chemicals that, when mixed, become the more toxic agent. These separate chemicals are called binary agents. This principle can be employed by a terrorist to lessen the danger of transporting dangerous chemicals which can be mixed at the point of delivery.

Military Nerve Agents

Sarin (GB), soman (GD), tabun (GA), and VX (V-agent) are the most widely known military neurotoxins. These agents are organophosphate compounds similar to the pesticides parathion and TEPP. See Table 3.1.

G (German) agents (GA, GB, and GD) are volatile and evaporate slightly faster than water. This makes them very dangerous as inhalation hazards. Terrorists can use a heat source with these chemicals to increase their volatility, thus increasing the hazard. Both tabun (GA) and sarin (GB) are easily synthesized, making them the terrorist choice for military-type nerve agents. Soman is not easily formulated and is the most deadly of the G agents. The strong bond formed between acetylcholinesterase and soman is a strong covalent bond (as opposed to an ionic bond). This bond is considered "aged" (strongly formed) and irreversible within two minutes of exposure.

VX ("V" for venom) is not as volatile as G agents, evaporating only as rapidly as motor oil. This makes VX primarily a skin absorption hazard. For VX to become a respiratory hazard, it must be mechanically aerosolized or heated to increase volatility. Because of the viscosity of this agent, its effects last longer, causing injury or death days later.

All of the military nerve agents are odorless. But because terrorists will not care how pure these substances are, nonmilitary chemicals have an odor. Clandestine G agents have a fruity odor, while VX has a sulfur odor. The military agents are also pH balanced to allow easier storage and to assist in their breakdown once dispersed. Nonmilitary nerve agents, typically not pH balanced, are very acidic. Depending on the pH value of the acidic nerve agents, air purifying respirators (APRs, specialized filter masks) will tend to break down rapidly when exposed. Also, military personnel protective equipment (PPE), including APRs designed for war agents, may not work as well during terrorist attacks.

Table 3.1 Neurotoxin (Nerve Agent) Overview

	TWA – ppm
Sarin (GB), isopropyl methylphosphonofluoridate, *$C_4H_{10}FO_2P$* **Odor:** Fruity **Route of entry:** Primarily inhalation, skin absorption, all routes possible	*.00002 ppm*

Boiling point: 297°F
Vapor density: 4.86

Soman (GD), pinacolyl methylphosphonofluoridate, $(CH_3)(C_6H_{13}O)POF$	***.000004 ppm***

Odor: Camphor
Route of entry: Primarily inhalation, skin absorption, all routes possible
Boiling point: 225°F
Vapor density: 6.33

Tabun (GA), O-ethyl N, N-dimethylphosphoramidocyanidate, $G_5H_{11}N_2O_2P$	***.00001 ppm***

Odor: Fruity
Route of entry: Primarily inhalation, skin absorption, all routes possible
Boiling point: 475°F
Vapor density: 5.6

VX, O-ethyl S-(2-diisopropylamino) ethyl methylphosphonothiolate, $C_{11}H_{26}NO_2PS$	***.0000009 ppm***

Odor: Sulfur
Route of entry: Primarily skin absorption, inhalation, all routes possible
Boiling point: 568°F
Vapor density: 9.2

Physiology of Neurotoxin Poisoning

These agents work by binding with the enzyme acetylcholinesterase, allowing the neurotransmitter acetylcholine to overstimulate nerve pathways. The nerve pathways affected by acetylcholine are primarily located in the parasympathetic nervous system, but are also found in the central and somatic system. Signs and symptoms of the overstimulation include excessive salivation, urination, diarrhea, excessive mucous production, muscle fasciculation, and seizures. The acronym DUMBELS (diarrhea, urination, miosis, bronchospasm, emesis, lacrimation, salivation) abbreviates the most prevalent signs and symptoms. See Table 3.3. A more complete list of symptoms is found in Table 3.2.

Table 3.2 Symptoms of Nerve Agent Poisoning

Nose:	Rhinorrhea
Eyes:	Constricted pupils, lacrimation, conjunctivitis (dim/blurred vision, pain)
Respiratory:	Hypersecretion production, bronchospasms (dyspnea, cough)
Gastrointestinal:	Cramps, diarrhea, vomiting, hypermotility

(continued)

Table 3.2 Continued

Somatic:	Weakness, fasciculations, seizures
CNS:	Anxiety, restlessness, coma, respiratory and circulatory depression
Other:	Salivation, sweating

Physical Properties and Routes of Entry

For the most part, these nerve agents are clear, colorless, and odorless. They are all found in liquid form with low vapor pressure and viscosity ranging from water-like to motor oil-thick. The most volatile of the group is sarin, which can evaporate at about the rate of water. These agents can enter through all routes, but inhalation causes the most rapid effects. Because an enemy cannot guarantee that inhalation will occur, some nerve weapons have thickeners ensuring that the agents will remain on objects for long periods of time and, thus, be transmitted to the victim. Once on the victim, the agent is absorbed through the skin or even ingested through contaminated food sources.

An exposure to vapors usually generates symptoms ranging from mild to severe within seconds or minutes. Mild symptoms include those defined in the acronym DUMBELS. Severe symptoms include all of the mild symptoms, with the addition of loss of consciousness, seizures, and apnea. When there is a liquid exposure, the onset of symptoms is slower, ranging from five minutes to eighteen hours.

Table 3.3 Common Acronyms Used to Describe the Effects of Organophosphate Poisoning

D-Diarrhea	**S**-Salivation	**S**-Salivation
U-Urination	**L**-Lacrimation	**L**-Lacrimation
M-Miosis (pinpoint pupils)	**U**-Urination	**U**-Urination
B-Bronchospasms (wheezing on auscultation)	**D**-Defecation	**D**-Defecation
E-Emesis (vomiting)	**G**-Gastrointestinal	
L-Lacrimation (tearing)	**E**-Emesis	
S-Salivation (increased saliva production)		

Decontamination

The military has performed many studies of the decontamination processes for nerve agents. Water, both fresh and sea, has the ability to decontaminate nerve agents through a process of hydrolysis. The process works well, but when thickening substances are added to the nerve agents, they are not easily soluble and hydrolysis occurs slowly. The addition of an alkaline soap has increased the efficiency of the decontamination process. Ideally, a sodium hypochlorite solution (household bleach) works best, causing oxidation of the nerve agent, which deactivates it. Refer to Chapter 8 for specific instructions on decontamination.

Treatment

The first treatment is to provide an open airway and support ventilation. Because patients are hypersecreting, the airways frequently become blocked and must be suctioned to ensure airway patency. Oxygenation should be accomplished before administering atropine to avoid possibly fatal cardiac dysrhythmias.

Antidotes for nerve agent poisoning include atropine IM or preferably IV, followed by pralidoxime chloride (protopam chloride, 2-Pam). Atropine is administered in high doses under close cardiac and vital sign monitoring. (This will not be possible if treating a large number of patients.)

Atropine blocks the effects of acetylcholine while the body metabolizes the organophosphate naturally. Atropinization must be maintained until all the absorbed organophosphate has been metabolized and the body again produces sufficient quantities of acetylcholinesterase. The treatment could last from days to weeks, necessitating the use of huge quantities of atropine. The normal dosage is 2–5 mg every five minutes until the mucous membranes dry. The military provides this to troops in the form of a kit (Mark I) containing 2 mg of atropine and 600 mg of pralidoxime for self-injection into intramuscular tissue.

Pralidoxime is the next drug indicated. The use of pralidoxime has three desirable results: frees and reactivates acetylcholinesterase; detoxifies the nerve agent; and has anticholinergic (atropine-like) effects.

Valium should be considered any time somatic system toxicity is noted (muscle fasciculations, seizures). If inhalation of the nerve agent occurred shortly after exposure it may be necessary to use the Valium before muscular involvement is noted.

■ NERVE AGENT—FIELD TREATMENT SUMMARY ■

Basic Treatment

1. Remove victim from the environment.
2. Decontaminate immediately using appropriate solutions.
3. Maintain an open airway.
4. Administer high-flow oxygen ASAP.
5. Continue with decontamination.
6. Consider advanced treatment if available and prepare for transport.
7. Monitor patient during transport. Suction if necessary.
8. Consider secondary contamination.
9. Rescuers should have appropriate level of protection.

Advanced Treatment

1. Follow procedures as identified in the basic treatment.
2. Evaluate oxygenation, high-liter flow before Advanced Life Support (ALS).
3. Give 2–5 mg atropine IVP every 5 minutes.
4. Give 1 gm pralidoxime over 2 minutes.
5. Give Valium for seizures.
6. Monitor patient for respiratory difficulty. Suction if necessary. Prepare for transport.

CHEMICAL ASPHYXIANTS (BLOOD AGENTS)

Chemical asphyxiants, those used commercially and those produced as by-products of industry, can be easily obtained by terrorists to inflict harm. The most commonly used chemical asphyxiant is cyanide. Cyanide is used for heat treating and plating, fumigation, and chemical synthesis in the production of plastics. It is found as a gas (hydrogen cyanide, AC), as a solid (cyanide salt), or as a liquid, and is a common component of many compounds containing carbon and nitrogen. Other common chemical asphyxiants in this group include hydrogen sulfide, carbon monoxide, and organic nitrogen compounds (nitrates and nitrites).

Military Blood Agents

Military chemical asphyxiants consist of two chemicals, hydrogen cyanide (AC) and cyanogen chloride (CK). These agents are identical to their civilian counterparts used in industry. For this reason, terrorists may choose these agents to inflict harm. Cyanide has a long history of use for both executions and homicides. For example, Jim Jones chose it to kill 900 people at Jonestown, and during an incident when Tylenol tablets were tainted with cyanide, others lost their lives. Even the chemist credited with the discovery of cyanide, Karl Scheele, died in a laboratory accident involving this chemical. See Table 3.4.

Table 3.4 Chemical Asphyxiant Overview

	TWA – ppm
Hydrogen cyanide (AC), HCN	*4.7 ppm*
Odor: Bitter almonds or peach kernels	
Route of entry: Primarily inhalation, skin absorption if liquid, all routes possible	
Boiling point: 79°F	
Vapor density: 0.93	
Cyanogen chloride (CK), CNCl	*.3 ppm*
Odor: Faint bitter almonds or peach kernels	
Route of entry: Primarily inhalation, if in liquid form (<55°F) can be skin absorbed	
Boiling point: 55°F	
Vapor density: 2.1	

AC is a liquid at less than 79°F, but vaporizes quickly. Because its vapor pressure is less than one, the gas rises and dissipates rapidly. For this reason, AC does not remain in surrounding areas. On the other hand CK is a liquid at less than 55°F, so in higher temperatures it is a gas. Its vapor density is 2.1, making it twice as heavy as air. It thus has a tendency to linger in low-lying areas, possibly inflicting harm for longer periods of time.

Physiology of Chemical Asphyxiants

Chemical asphyxiants inflict harm in one of two ways, depending on the chemical. One way is to inhibit the hemoglobin's ability to carry oxygen. This is related to carbon monoxide (CO) and nitrate/nitrite poisoning. The other is to interfere with a cell's ability to use oxygen. Both cyanide and hydrogen sulfide work in this way.

Hemoglobin is the complex molecule that contains iron atoms capable of bonding with oxygen for transport to a cell. This molecule is found as part of the red blood cell in the circulatory system. When the red cells flow by the oxygenated lungs, they pick up huge amounts of oxygen for transport to the body cells, where it is distributed as needed. Chemicals like CO and nitrates/nitrites interfere with this process of transportation. Depending on the amount of hemoglobin affected, the symptoms can range from headache to death. See Table 3.5.

Table 3.5 Signs and Symptoms of Carbon Monoxide and Nitrate/Nitrite Poisoning

% of Hemoglobin Disrupted	*Signs and Symptoms*
1–10%	No symptoms
10–20%	Headache
20–30%	More severe headache and difficulty reasoning
30–40%	Very severe headache, nausea/vomiting, dim vision
40–60%	Coma and convulsions
>60%	Cardiovascular collapse, respiratory failure, death

Carbon monoxide

CO actually bonds with the iron in the hemoglobin and does not release at the cell as oxygen would. Instead, it ties up the bonding sites normally used by oxygen, causing hypoxia at the cellular level, which leads to asphyxia and death. The treatment for CO poisoning consists of supplying 100% oxygen using positive pressure ventilations and definitive hyperbaric treatments.

Nitrates/nitrites

Nitrate/nitrite poisoning also involves hemoglobin, but in a different way. These chemicals actually change the bonding capability of the iron atom, causing it not to bond with oxygen. The hemoglobin that results is called methemoglobin. In this case, the methemoglobin cannot carry oxygen because it is not attracted to it. Interestingly, methemoglobin is attracted to cyanide and is used as an antidote to cyanide poisoning. Methemoglobin can be changed back to hemoglobin with the methemoglobin antidote called methylene blue, commonly found in hospital pharmacies and on some hazardous materials units.

In either of these cases of poisoning (CO or nitrate/nitrite), death is imminent once 60–70% of the hemoglobin is out of service for the transportation of oxygen. However, the body can stand various levels of hypoxia with full recovery. If treatment is attempted on an unconscious victim, 100% oxygen is necessary with positive pressure ventilations if a successful resuscitation is expected.

Cyanide (AC and CK) and hydrogen sulfide

These agents work by inhibiting the ability of cells to use oxygen. Following an exposure, the cyanide or sulfide ion enters the cells, binding with the enzyme cytochrome oxidase. This enzyme, in its original form, is necessary for cellular respiration (the use of oxygen to convert glucose to energy). By binding to cytochrome oxidase, these ions cause a paralysis of the cells' ability to carry out aerobic metabolism. Without cytochrome oxidase, a cell cannot use oxygen from the blood stream. The process eventually causes the cells to attempt to function under an anaerobic metabolism, and this inadequate anaerobic state ultimately causes decreased cellular energy production, metabolic acidosis, cellular suffocation, and death. The half-life of cyanide in the body is only about an hour, but during a true exposure, death takes place well before the body starts to detoxify or excrete the chemical. See Table 3.6.

The ability to detect the chemical's bitter almond smell is a sex-linked recessive trait of only 60–80% of the general population. The remaining population cannot detect the odor due to a sensory deficit, greater in men than women by a ratio of 3 to 1.

Table 3.6 Signs and Symptoms of Cyanide (Hydrogen Cyanide and Cyanogen Chloride) and Hydrogen Sulfide Poisoning

Respiratory System Effects	
Early	***Late***
Tachypnea (rapid breathing)	Decreased respiratory rate
Hyperpnea (deep breathing)	Respiratory depression
Dyspnea (shortness of breath)	Apnea and death
Cardiovascular System Effects	
Early	***Late***
Flushing (circulatory system fully oxygenated)	Hypotension
Hypertension	Acidosis
Reflex bradycardia	Tachycardia
Arterioventricular (AV) nodal or intraventricular rhythms	EKG-ST segment changes

Physical Properties and Routes of Entry of Cyanide Agents

Cyanide is one of the most rapidly acting poisons. It gains access to the body most often through inhalation, but can also be ingested and absorbed through the skin and mucous membranes. It causes death within minutes to hours, depending on its concentration and route of entry as well as the exposure time and activity level of the victim. The speed at which cyanide gas works is evidenced by how rapidly (usually within a minute) a death row prisoner dies during a gas chamber execution.

The patient will present with a wide variety of symptoms because cyanide poisoning affects virtually all of the cells in the body. The most sensitive target organ is the central nervous system, where the urgent need for oxygen is first sensed. Early effects can include headache, restlessness, dizziness, vertigo, agitation, and confusion. Later signs are seizures and coma.

Because both of these agents rapidly vaporize, they readily mix with air and are, for the most part, nonpersistent. Since respiratory system exposure is the most common route of entry, a cartridge mask or self-contained breathing apparatus is needed if a rescue is attempted in a contaminated atmosphere. CK can break down charcoal filters if exposure is for a long duration or repeated. Frequent changes of filters are recommended.

Decontamination

If decontamination is needed (exposed to liquid or solid), a mild bleach solution is recommended. Exposure to the gas will warrant the removal of clothing prior to treatment.

Treatment

Since respiratory arrest develops quickly, establish a good patent airway. Then, as quickly as possible, begin advanced treatment using a cyanide antidote kit. The kit contains amyl nitrite, sodium nitrite, and sodium thiosulfate.

The nitrites (amyl nitrite and sodium nitrite) convert hemoglobin into methemoglobin. Methemoglobin competes with cytochrome oxidase for the cyanide ion, actually attracting the cyanide away from the cytochrome oxidase, which frees the cytochrome oxidase to participate again in aerobic cellular metabolism. Last, infuse sodium thiosulfate, which acts as a cleanup agent by changing the remaining cyanide into a relatively harmless substance, thiocyanate.

The patient who survives an initial exposure must be closely monitored and admitted into the hospital. This precaution is necessary because of the complications of acidosis, pulmonary edema, dysrhythmias, and neurologic deficits that may appear many hours after an exposure.

■ CHEMICAL ASPHYXIANT—FIELD TREATMENT SUMMARY ■

Basic Treatment

1. Remove victim from environment.
2. Decontaminate using appropriate solutions.
3. Maintain an open airway.
4. Use high-flow oxygenation.
5. Prepare for transport.
6. Consider secondary contamination.
7. Rescuers should have appropriate level of protection.
8. Consider advanced treatment if available.

Advanced Treatment

1. Procedures are as identified in the basic treatment.
2. Use amyl nitrite perles inhaled for 15–30 seconds every minute until IV is established.
3. If patient is unconscious, drop perles into the bag valve mask and ventilate.
4. Administer sodium nitrite, 300 mg per 10 cc IV slowly. Both nitrites will cause vasodilation, so expect a drop in blood pressure. This can normally be corrected with positioning.

(continued)

Advanced Treatment (continued)

5. Administer sodium thiosulfate, 50 ml of a 25% solution IV over 10 minutes. This can be accomplished using a second IV or infused through a piggyback bag. (Do not use step 5 for H_2S.)
6. Monitor patient during transport.
7. Hyperbaric oxygen treatment is beneficial but not mandatory.

Children's Dosage

1. Amyl nitrite remains the same as for an adult.
2. Use sodium nitrite, 0.33 ml/kg (10mg/kg of a 3% solution).
3. Use sodium thiosulfate, 1.65 ml/kg of a 25% solution.
4. Hyperbaric oxygen treatment is beneficial but not mandatory.

RESPIRATORY IRRITANTS (CHOKING AGENTS)

Strong respiratory irritants have a long history of use by military forces. World War I documents are filled with incidents of chemical warfare involving chlorine (Cl) and phosgene (CG) gases. Today, these gases remain in military arsenals around the world. When discussing terrorism, we cannot discount these agents.

Many communities use Cl gas in its pure form for chlorinating drinking water. It is also used as an antimold and fungicide agent. Compounds containing chlorine are even used for chlorinating home swimming pools and cleaning toilets and showers. There is no doubt that this chemical is easily available.

CG, although not as common as chlorine, is found in industry and used for organic synthesis during the production of polyurethane, insecticides, and dyes. It is also a by-product of burning Freon, which has led to many injuries among firefighters.

Although not thought of as a military agent, anhydrous ammonia also fits into this category because of its ability to cause severe respiratory irritation and injury. Anhydrous ammonia is a common industrial refrigerant and is also used in blueprinting. Again, it is easily available to anyone. See Table 3.7.

Table 3.7 Respiratory Irritants Overview

	TWA – ppm
Phosgene (CG), carbonyl chloride, $COCl_2$	*.1 ppm*
Odor: Mown hay, sweet	
Route of entry: Inhalation	
Boiling point: 45°F	
Vapor density: 3.4	
Chlorine (Cl), Cl_2	*.5 ppm*
Odor: Bleach	
Route of entry: Inhalation, also causes irritation where the skin is moist	
Boiling point: -29°F	
Vapor density: 2.49	

Nonmilitary, Commonly-found Respiratory Irritant

Anhydrous ammonia, NH_3 ***25 ppm***

Odor: Sharp, intensely irritating
Route of entry: Inhalation
Boiling point: -27°F
Vapor density: 0.58

Military Choking Agents

CG and Cl are typically stored as liquids, but rapidly become gases once released into the atmosphere. Their expansion ratios allow them to be transported in small containers, and, once released, they fill a large area.

These agents were used on the battlefield to incapacitate an enemy force, so that it could be overrun by advancing troops. This strategy worked well, because both of these gases dispersed rapidly into the environment, leaving no contaminated objects behind to cause injury.

Once exposed, victims are overcome with severe, uncontrollable coughing, gagging, and tightness in the chest. Bronchospasms and laryngeal spasms are common, causing apnea and unconsciousness. Other injuries include tissue sloughing, localized edema, and pulmonary edema, all contributing to obstruction of the airways. CG, in particular, is not easily soluble in water and advances directly into the alveoli, where tissue destruction is severe. The breakdown of these cells allows fluid from the bloodstream to advance into the airways, a condition called noncardiogenic pulmonary edema, difficult to treat. The victim of a severe exposure literally drowns in his own body fluids. When the injury is not fatal, the victim is left with destroyed lung tissue and a lifetime of respiratory disease ranging from chronic obstructed pulmonary disease to chronic pneumonia.

Physiology of Respiratory Irritant Injury

Injuries to the upper respiratory areas are usually a result of exposure to water-soluble chemicals that readily dissolve into the moisture-coated airways. In the case of chlorine, this results in the production of hydrochloric acid and chemical burns at the site. Ammonia in reaction with water forms an alkali, which causes a severe, long-lasting injury. Laryngeal edema and laryngeal spasms should be expected and treated aggressively.

Injuries to the lower regions of the respiratory tract usually result when the chemical inhaled is not water soluble, is in a high concentration, or is inhaled over an extended period of time. This deeper injury causes swelling of the finer bronchioles, sloughing of damaged tissue, and damage to the alveoli. Cilia that may be damaged are unable to rid the fine bronchioles of the sloughed cells and increased mucous production caused by the damaged airways. The lower airway obstruction caused by this exposure adds to the complexity of the injury and increases the chance of death.

Noncardiogenic pulmonary edema is the result of a chemical irritant reaching the alveoli and causing damage. The damage interferes with the alveoli's natural ability to keep fluids out of the alveolar space, which results in fluids filling the injured alveoli and fine bronchioles, eventually advancing into the upper airways.

If the exposure causes severe alveolar damage, the end result will be death. See Table 3.8.

Table 3.8 Signs and Symptoms of Respiratory Irritant Injury

1. If the chemical is **highly soluble in water**, the typical injury will involve the upper respiratory system. Bronchospasms, laryngeal spasms, localized chemical burns, and irritation are likely to occur. With auscultation, wheezing will be noted due to upper-airway swelling and occlusion. This is the common injury found with exposure to anhydrous ammonia and, to some degree, chlorine.
2. If the chemical has **poor solubility in water**, the injury will be deeper in the airways and will probably affect the fine bronchioles and alveoli. Auscultation will reveal rales and the probability of noncardiogenic pulmonary edema.

Physical Properties and Routes of Entry

Military chemical irritants are heavier than air and are able to seek out low-lying areas and persist for long periods of time. This was the case during World War I, when troops would dive for cover from gunfire, only to choke in their foxholes. In the civilian world, hazardous materials responders have been dealing with these chemicals for years. The typical responder would point hose lines toward the cloud of gas and disperse it using a combination of air currents and water absorption.

All of these agents are referred to as respiratory irritants, but the respiratory system is not the only body surface affected. If the irritant is water soluble (ammonia and chlorine), it will cause moist skin and eyes to become inflamed and burn. Both of these effects are unpleasant and could put troops or civilians out of action. Even at low concentrations, the release of these chemicals into a crowd could send multitudes of patients to the hospital, triggering a panic scenario in line with the needs of a terrorist.

Decontamination

Decontamination with a solution is normally not necessary. All of these chemicals are gases and will disperse rapidly into the environment. Clothing removal may be necessary if the patient is to be placed in a closed ambulance for transport.

Treatment

The treatment for respiratory irritant exposure is twofold. The first part is to open the airways to allow the free movement of air into the alveoli. The second part of the treatment is to reduce the fluid in the alveoli to allow gas exchange with the blood. Getting oxygen to the alveoli is vitally important. If oxygen is available, there is never a reason to withhold it during treatment of these exposed victims.

In upper respiratory injury, oxygen must pass through narrowed passageways to gain access to the lower system. Bronchodilators like alupent and albuterol given in an updraft will provide some dilation of the airways; brethine and epinephrine given subcutaneously may also assist in making the bronchioles larger to allow for the passage of air and oxygen. However, care must be taken with any of these drugs, because their side effects are high blood pressure and rapid heart rate. These are also common with hypoxia.

When lower airway injury results, the outcome is usually pulmonary edema (PE). Chemically induced (noncardiogenic) pulmonary edema is difficult to treat and may involve a range of actions. The normal field treatment of this injury is to decrease pulmonary pressure using nitroglycerine, morphine, and lasix. But these measures will not reduce chemically induced pulmonary edema, which does not respond to reduced pulmonary pressure. Instead, the only field treatment that will assist patients with this condition is positive pressure ventilations using a positive end expiratory pressure (PEEP) valve. Providing positive pressure during inhalation, exhalation, and at the end of the respiratory cycle will assist the movement of fluid back into the vascular space and limit the influx of new fluid.

■ RESPIRATORY IRRITANT—FIELD TREATMENT SUMMARY ■

Basic Treatment

1. Remove victim from environment.
2. Decontaminate immediately using the appropriate solutions.
3. Maintain an open airway.
4. Use high-flow oxygenation.
5. Consider secondary contamination. Rescuers should have the appropriate level of PPE for activity.
6. Prepare for transport.
7. Monitor patient's respiratory status carefully.

Advanced Treatment

1. Use procedures as identified in the basic treatment.
2. Upper airways can be dilated using updrafts of alupent or albuterol.
3. Further dilation can be accomplished using brethine or epinephrine subcutaneously.
4. **Lower airway injury usually results in pulmonary edema. Normal treatments for cardiogenic pulmonary edema will probably not work, so positive pressure ventilations using a PEEP valve will treat the injury.**
5. Monitor patient during transport.

■ SKIN AND EYE IRRITANTS (BLISTER AGENTS, VESICANTS)

These chemicals were originally developed by the military because enemy troops could protect themselves from respiratory irritants with masks. Instead, skin and eye irritants could affect the respiratory system and additionally unprotected skin. These irritants are harmful in liquid form, but vaporize to become airborne contaminants as well. Most date back to World War I; some have been further refined through the years to become even more efficient.

Many chemicals in industry are capable of causing skin irritation, but none to the degree that military blister agents can. It is for this reason that any discussion about skin irritants in this book will center on military agents.

Military Blister Agents (Vesicants)

Three types of blister agents are primarily used by the military. These agents include mustard (H) and related variations of mustard (HD, HN, and HT), phosgene oxime (CX), and lewisite (L). For the most part, the agents are liquids that vaporize slowly causing an inhalation hazard. Skin and eye exposure is the most common effect that results from direct contact with the liquid. CX is normally a solid that will also vaporize into a respiratory hazard.

H was developed during World War I and has continued to be a major chemical warfare agent since that time. This agent was reportedly used in 1960 by Egypt against Yemen, and again during the 1980s as a weapon between Iran and Iraq. The United States still maintains stockpiles of this agent in both Colorado and Utah but is working to incinerate the weapons. See Table 3.9.

Table 3.9 Skin Irritant Overview

	TWA – ppm
Mustard (H, HD), bis (2-chlorethyl) sulfide, $(ClCH_2CH_2)_2S$ or $C_4H_6Cl_2S$	*.0003-.0004 ppm*
Odor: Garlic	
Route of entry: Primarily skin absorption; eye and respiratory also possible	
Boiling point: 442°F	
Vapor density: 5.4	
Phosgene oxime (CX), dichloroformoxime, $CHCl_2NO$	*.086 ppm*
Odor: Penetrating, irritating odor	
Route of entry: Primarily skin absorption; eye and respiratory also possible	
Boiling point: 264°F	
Vapor density: 3.9	
Lewisite (L), dichloro-(2-chlorovinyl) arsine, $ClCHCHAsCl_2$	*.0003 ppm*
Odor: Geraniums	
Route of entry: Primarily skin absorption; eye and respiratory also possible	
Boiling point: 374°F	
Vapor density: 7.2	

Physiology of Blister Agent Exposure

Strong irritants, these agents are capable of causing extreme pain and large blisters on contact. If the vapors are inhaled, the lung tissue will form large obstructing blisters. Once the blisters break, a large open wound results that allows the establishment of overwhelming infections, a condition that will eventually cause death.

After H gains access to the body, it cycles in the extracellular water forming an extremely reactive substance binding to both intracellular and extracellular enzymes and proteins. Once exposed, a victim may manifest a latent period when no symptoms are present. Later, between two and twenty-four hours, the reaction appears with the formation of blisters on the skin, necrosis of the mucosa of the airways with later involvement of the airway musculature, and severe irritation to the eyes, including swelling in the cornea and related tissues that leads to permanent scarring.

L acts similarly to H, but has additional effects that are systemic. Symptoms beyond the blistering effect may include pulmonary edema, diarrhea, vomiting, weakness, and low blood pressure. The irritation liquid L causes to the eyes is devastating. If not decontaminated within one minute, damage will probably be permanent. See additional signs and symptoms in Table 3.10.

Table 3.10 Signs and Symptoms of Blister Agent Exposure

Mustard (H)	*Lewisite (L)*	*Phosgene Oxime (CX)*
Skin	**Skin**	**Skin**
symptoms in 2–24 hours	**symptoms immediately**	**symptoms immediately**
Erythema	Dead grayish	Blanching
Pruritis	epithelium	Erythematous ring,
Burning	Blisters over 12-18 hrs.	30 seconds
Blisters	**Pulmonary**	Wheals in 30 minutes
Pulmonary	Extreme,	Pain persists for days
Nose, sinus,	immediate irritation	**Pulmonary**
pharynx burning	Pseudomembrane	Immediate irritation
Epistaxis	formation	Pulmonary edema
Pseudomembrane	Pulmonary edema	**Ocular**
formation	**Ocular**	Pain
Airway obstruction	Pain	Blepharospasm
Ocular	Blepharospasm	Conjunctival edema
Irritation	Conjunctival edema	Iritis and
Photophobia	Iritis and	corneal damage
Blepharospasm	corneal damage	
Pain		

Physical Properties and Routes of Entry

H has a freezing point of 57°F, so it solidifies at temperatures less than this. For this reason, pure mustard may not be a good choice in colder climates. H also vaporizes slowly, making it primarily a skin contact hazard. At temperatures greater than 100°F, it will vaporize rapidly enough to be a respiratory hazard. Mixing H with some other agent like L lowers its freezing temperature and allows the use of this chemical in colder climates.

Both L and CX vaporize more readily than H, making them more of a respiratory hazard. All of the blister agents have vapor densities much greater than air, allowing them to stay near the ground and not dissipate quickly. L has a vapor density of 7.2, which allows it to persist in low-lying areas for long periods of time.

Decontamination

Decontamination for all of the blister agents must be immediate. Each one harms tissues on contact. H differs from the other agents, as it does not produce symptoms for several hours, leaving the victim without a clue that an exposure has taken place. L and CX both cause irritation almost immediately, which alerts the victim to exposure and allows earlier decontamination. If a victim notices liquid on his/her skin, he/she should blot off the agent, taking care not to do this with contaminated material. If the agent is wiped instead of blotted, it will be spread the length of the wiped area extending the injury.

The decontamination solution of choice is a chloramine solution. CX is adequately decontaminated with water, but a more efficient decontamination includes soap and water followed by a chloramine or bleach solution, followed again by soap and water.

■ SKIN IRRITANT TREATMENT ■

There is no field treatment for blister agents beyond good decontamination. It is important to know that the blisters formed by these agents do not pose a significant secondary contamination danger. The fluid in the blisters is not contaminated with the agent, but typical blood and body fluid precautions should be exercised.

■ RIOT CONTROL AGENTS (LACRIMATING AGENTS)

Riot control agents are used to incapacitate enemies and make them unable to fight. They are not intended to cause mortal injury and have only rarely caused severe lasting injury. Civilian use of these agents includes riot control and self-protection. Chemical antipersonnel weapons have gained popularity with both the general public and law enforcement, because they can be used to subdue persons without the use of extraordinary physical force. These chemical sprays offer a nonlethal form of protection that causes temporary extreme discomfort. Generally, there are three versions of these sprays available: chloracetephenone (CN), orthochlorobenzalmalonitrile (CS), and the most popular civilian agent, oleoresin capsicum (OC). Additional chemical names are found in Table 3.11.

Table 3.11 Synonyms for Typical Riot Agents

Chloracetephenone (CN)
Phenacyl chloride, alpha-chloracetephenone, omega-chloroacetephenone, chloromethyl phenyl ketone, and phenyl chloromethyl ketone

Orthochlorobenzalmalonitrile (CS)
O-chlorobenzylidene malonitrile, OCMB, and military tear gas

Oleoresin capsicum (OC)
Pepper spray, civilian tear gas

CN and CS

The effects from CN begin in 1–3 seconds and are characterized by extreme irritation to the eyes, causing burning and tearing. Irritation to the skin is also common, because the crystals stick to moist skin, causing burning and itching at the point of contact. CN also causes upper respiratory irritation. These effects last 10–30 minutes.

The symptoms of CS start in about 3–7 seconds and last 10–30 minutes. The effects reported are stinging of the skin, especially in moist areas, and intense eye irritation with profuse tearing and blepharospasms. The burning also affects the nose and upper respiratory system. Some victims panic, due to shortness of breath and chest tightness; they describe its effects as being ten times worse than CN. Police agencies still use this irritant for crowd dispersal.

Both CN and CS are submicron (less than 1 micron) particles. They are extremely light and are carried to the target area in a carrier solution that evaporates quickly, dispersing the agent. Because of their light, fine particles, both of these chemicals are capable of cross contamination between the victim and emergency response personnel. Their submicron size may allow these irritating particles to gain access to the lungs, fine bronchioles, and alveoli, causing injury in these places.

OC

OC has become the safest and most popular of the chemical agents. It is found in police aerosol sprays and over-the-counter agents. It is a non-water-soluble agent prepared from an extract of the cayenne pepper plant. When contact with the eyes occurs, the effects of OC are almost immediately apparent. OC is not a submicron particle, so access to the lower respiratory system is limited. Contact with OC causes immediate nerve-ending stimulation, but not irritation. The condition lasts 10–30 minutes and usually has no lasting effect.

■ TREATMENT FOR EXPOSURE TO RIOT CONTROL AGENTS ■

Less than 1% of the exposed victims will have symptoms severe enough to need medical care. As mentioned earlier, the symptoms generally last about 30 minutes and disappear without lasting injury.

There is no known antidote for these irritants, so medical care is centered on relief of the symptoms. In the case of eye irritation caused from exposure, relief from symptoms can be accomplished with a topical anesthetic agent like alcaine, ponticaine, or opthalmicaine. If sensitivity to these chemicals causes bronchospasms, typical bronchodilating drugs may be used. Updrafts of albuterol or alupent will usually control bronchospasms caused by exposure to these agents.

Decontamination

Decontamination should consist of removing the clothing and washing exposed areas with soap and water. In the case of OC exposure, the agent is non-water-soluble, so its effect on mucous membranes will last until it is detoxified by the body, not washed away through irrigation. This detoxification takes place in about 30 minutes, which is the duration of OC's effect. Irrigation of the eyes is usually not necessary but will, in itself, relieve the burning pain. Except for the blepharospasm, which limits ability to open the eyes, vision is unaffected.

CHEMICAL VARIATIONS

Each chemical discussed in this chapter can be varied by adding or subtracting a functional group or separate molecule. For example, hydrogen cyanide or hydrocyanic acid (AC) is one of the most rapidly acting poisons. Cyanide enters the cell, binding with cytochrome oxidase and causing paralysis of the cell's ability to carry out aerobic metabolism. A variation of hydrocyanic acid is Zyklon B, a powder. Easily placed into a fume and dispersed over a large area, its overall effectiveness is shorter compared to AC, and it is extremely volatile and toxic. Table 3.12 is a brief list of other possible configurations.

Table 3.12 Variations of Typical Military Agents

Nerve Agents

VX is the standard nerve agent. Variations include VE, VG, and VS. These are all colorless and odorless. They do not freeze and evaporate slowly at normal ambient temperatures, so they persist in the environment. Aerosolized droplets once on the skin are rapidly absorbed causing injury. Action is extremely fast compared to the G agents.

Blood Agents

Arsine trihydride (SA) is a colorless, nonirritating gas with a garlic odor. Fatality rate is 25% or greater. It causes renal failure and pulmonary edema, which are secondary to the primary injury of red blood cell destruction by hemolysis. Fifty ppm can produce death; TWA .05 ppm.

Stibine gas (antimony hydride) has a TLV of .1 ppm. Exposure is common through inhalation, with injuries similar to SA.

Choking Agents

Polytrafluoroethene (PTFE) has no reported odor and may also be referred to as PFIB or perfluoroisobutene. Usually, the patient reports difficulty in breathing with or without substernal chest pain or discomfort, coughing that is nonproductive, and a headache. This occurs in the first 1–2 hours. Symptoms worsen over the next 8–12 hours, with pulmonary edema 8–48 hours after exposure.

Disphosgene (DP) is similar to its cousin CG. DP converts in the body to CG causing the toxic effect. Trichloromethyl chloroformate is its chemical name, and it is extremely volatile under normal conditions. Evaporation is fast, causing it to disseminate rapidly. However, because of this, it is difficult to weaponize.

Vesicants

H has many variations. All cause the same type of injuries, some slower some faster, depending on temperature, concentration, and chemical makeup.

Mustard-Lewisite	HL
Phenyldichloroarsine	PD
Ethyldichloroarsine	ED
Methyldichloroarsine	MD
Levinstein mustard	H

As you can see, the list is long. Remember that common hazardous materials can also be used and are probably easier and less technical for terrorists. Consider these scenarios: a small bomb set off at the local gas station or wastewater facility, releasing fuel or chlorine, respectively, over a populated area, or a bullet penetrating a LPG tanker moving through a community. The possibilities are endless.

4

BIOLOGICAL AGENTS

OVERVIEW

The thought of being infected by a deadly disease or poisoned by a biological toxin is truly a frightful one. This fear may stimulate a terrorist to choose a biological weapon. Unfortunately, these agents are not difficult to cultivate; they are surprisingly easy for someone with a very limited knowledge of microbiology to produce. Although the threat of this type of attack may seem remote, it is exactly what would bring international attention to a radical organization.

Organized research for biological warfare gained momentum during World War II. Prior to this time, biological agents were used to inflict harm on individuals, but were not, for the most part, cultivated or produced in mass quantities for war. For example, plague-ridden corpses were flung over fortress walls in the fourteenth century and smallpox-ridden blankets were given to the American Indians during the French and Indian War of 1754. True biological warfare research, however, did not take place until much later.

Biological agents are made from a variety of microorganisms and biological toxins. Biological toxins are chemical compounds poisonous to humans produced by plants, animals, or microbes. Microorganisms are generally living viruses or bacteria that have the ability to establish deadly infections in their victims. Although many of these organisms are recognized as military-type weapons, many others can be cultivated and introduced into the environment with the intention of inflicting harm on a targeted civilian population.

Bacteria

Bacteria are single-celled microorganisms plant-like in structure. They vary in size from approximately one half of a micron to ten microns. They can be either spherical (cocci) or rod shaped (bacilli). Bacterial agents include living cells of *Bacillus anthracis* (anthrax), cholera, *Yersinia pestis* (plague), *Francisella tularensis* (tularemia), Q fever, and salmonellae. These microorganisms can be grown in artificial media; many have the ability to spore (become seed-like) and live for long periods of time without infecting tissue.

Viruses

Viruses are smaller than most bacteria and live on or within other cells, using the host cells' machinery for metabolism. A viral infection is the result of destruction to host cells by the intracellular parasitic action of the virus. Viruses cannot be grown in artificial media, only in media that contain living host cells. Each virus needs a particular type of host cell, making the production of viruses for warfare or terrorist use complicated and expensive. For this reason, it is probably unlikely that low-budget organizations or private individuals would use viruses to inflict harm on a target population.

Military forces from many countries have experimented with virus use for weapons. This chapter will provide an overview of those viruses considered most threatening, including variola virus (smallpox), Venezuelan equine encephalitis (VEE), and viral hemorrhagic fever (VHF).

Biological Toxins

Biological toxins are toxic substances originating in animals, plants, or microbes, and are more toxic than most chemicals used and produced in industry. Since these toxins are not volatile, they are generally not suitable for the battlefield. Instead, they are used for contaminating food sources, water supplies, and specific targeted individuals. Toxins that have been considered for military use include botulinum toxins (botulism), staphylococcal enterotoxin B (SEB), ricin, and tricholthecene mycotoxins (T2s).

BACTERIAL AGENTS

Anthrax *(Bacillus anthracis)*

Anthrax (splenic fever, woolsorters' disease, malignant pustule) has long been the biological weapon of choice. It was first prepared as a weapon in the United States in the 1950s and continued to be produced until the biological weapon program was terminated. Anthrax is easily grown and can be kept almost indefinitely in the spore (dormant) form under the proper conditions. Although the United States no longer cultivates anthrax as a weapon, it is suspected that many other countries have biological weapons using this agent. It is reasonable to expect that terrorist organizations are either developing a weapon of this type or already have one at their disposal.

Anthrax can be delivered to a target as dust that can be inhaled. Furthermore, it can contaminate the environment and drinking water, causing disease days or weeks after it is disseminated.

Once anthrax bacteria are inhaled or ingested, they will incubate for approximately one to six days in the victim. Signs of anthrax poisoning include chest pain, cough, fatigue, and fever. More serious symptoms develop as the infection becomes worse and include shortness of breath, diaphoresis, and cyanosis, leading to shock and death.

Treatment includes high-dose antibiotics and may only be successful if the infection is caught in the early stages. Other supportive therapy, such as intubation and ventilation, may also be necessary. If an anthrax infection is suspected, precautions by responders should include HEPA-style masks and blood and body fluid protection.

■ SUMMARY ■

Anthrax **(Bacillus anthracis)**
Onset of symptoms: Symptoms appear in 1–6 days.
Types of symptoms: Chest pain, cough, fatigue, and fever progress to shortness of breath, diaphoresis, cyanosis, and death.
Types of dispersion: Delivered as a dust, it will contaminate the environment and drinking water.
Personal protection and decontamination: HEPA masks, blood and body fluid protection are necessary. Decontamination should be accomplished using a hypochlorite solution.

Cholera *(Vibrio cholerae)*

The bacteria responsible for the cholera infection gains entrance into the body through contaminated food and water sources. It is a disease that spreads rapidly when proper sewage precautions are not taken. For this reason, it has been investigated as a wartime biological agent. If an infection can be established in a field camp where sewage precautions are not taken, the disease can infect a high percentage of the troops. In countries where sewage disposal is not carefully monitored, cholera has infected and killed of thousands of people.

The cholera bacterium attaches to the tissue of the small intestine, causing an oversecretion of fluid. This overwhelms the large intestine's ability to absorb the fluid and leads to diarrhea and severe hypovolemia.

Signs of cholera begin within twelve to seventy-two hours of exposure and include vomiting, intestinal cramping, and headache. Five to 10 liters of fluid loss can be expected per day. If not aggressively treated, the fluid loss will lead to shock and death.

Treatment includes fluid replacement with electrolyte therapy. Antibiotics will shorten the duration of the infection and kill the infecting microorganisms. It is difficult to become infected through direct contact with an infected person, but precautions should be taken to reduce contact with body fluids. Hypochlorite solutions should be used as a decontaminating agent for material or equipment that has been in contact with body fluids.

■ SUMMARY ■

Cholera **(Vibrio cholerae)**
Onset of symptoms: Symptoms appear in 12–72 hours.
Types of symptoms: Vomiting, intestinal cramping, headache, and severe diarrhea occur.
Types of dispersion: It is disseminated through contaminated drinking water.
Personal protection and decontamination: Protection from blood and body fluids is necessary. Person-to-person transmission is not common, but contaminated areas should be washed with a hypochlorite solution.

Pneumonic/Bubonic Plague *(Yersinia pestis)*

Both pneumonic and bubonic plague are caused by the same bacteria, but their symptoms are different. Normally spread by rodents or fleas, the organisms could be introduced into the environment through aerosolized bacteria for wartime use. Japan has researched the feasibility of releasing plague-infected fleas as a means of dissemination.

Swollen lymph nodes (buboes) and fever characterize bubonic plague. The affected lymph nodes are most often those found in the groin, since infected fleas often bite the leg areas. The incubation period for a bubonic infection is between two and ten days.

Pneumonic plague's onset is usually faster and follows an incubation period of two to three days. Infection results from a respiratory exposure to *Yersinia pestis* spread from person to person and caused by an infected host coughing or sneezing. Symptoms of pneumonic plague include fever, chills, coughing, bloody sputum, and dyspnea and cyanosis. If left untreated, both bubonic and pneumonic infection can progress into a more serious septicemic form that attacks the central nervous system and other organs.

Treatment of plague involves early diagnosis and administration of antibiotics. Other supportive measures may be needed if breathing is impaired or other organ systems are involved. Personnel protection is necessary during treatment of these patients, since secretions and body fluids may be infectious if bubonic plague is the cause of illness. If pneumonic plague is the cause, strict respiratory protection is necessary, along with clothing protection, to prevent secondary contamination. Decontaminate any equipment using a hypochlorite solution.

■ SUMMARY ■

***Pneumonic/Bubonic plague* (Yersinia pestis)**

Onset of symptoms: Symptoms appear in 2–10 days, fewer for respiratory exposure.

Types of symptoms: Swollen lymph nodes in the groin and legs are most common if infected by fleas. Respiratory exposure causes coughing, sneezing, fever, chills, bloody sputum, cyanosis, systemic effects, and septicemia.

Tularemia *(Francisella tularensis)*

Tularemia was prepared as a weapon by the United States during the 1950s but was discontinued with the termination of the biological weapon program. It is reasonable to expect that other countries are still cultivating tularemia as a weapon.

Tularemia is also known as rabbit fever or deer fly fever. Blood and body fluids of an infected person or animal, or the bite of an infected deer fly, tick, or mosquito, transmit the disease. Inhalation of aerosolized bacteria would initiate a typhoidal tularemia infection with respiratory symptoms that appeared in two to ten days. Tulaemia persists for weeks in water, soil, or animal hides. Because the bacteria is resistant to freezing, it can persist in frozen rabbit meat for years.

Signs and symptoms of an infection include local ulcerations, swollen lymph nodes, fever, chills, and headache. Typhoidal symptoms include fever, headache, substernal discomfort, and a nonproductive cough. Even untreated, the death rate for this infection is about 5%.

Excellent results have been obtained with treatment that includes antibiotic therapy. Secondary infection is unusual, so strict isolation is not needed. Only typical personal protection against secretion and lesions is required.

■ **SUMMARY** ■

Tularemia **(Francisella tularensis)**

Onset of symptoms: When the organism is inhaled, symptoms appear in 2–10 days.

Types of symptoms: Symptoms include local ulceration, swollen lymph nodes, fever, chills, and headache.

Types of dispersion: The organism is dispersed in either wet or dry form as an aerosol.

Personal protection and decontamination: Blood and lesion protection is necessary, but patient isolation is not. Decontamination is accomplished using hypochlorite solution.

Q Fever *(Coxiella burnietii rickettsia)*

Q fever (query fever) is another weapon previously kept in U.S. arsenals. It occurs naturally as an infection found in sheep, cattle, and goats, and is an occupational hazard in industries involving these livestock. Q fever is a rickettsial agent disseminated through inhalation of infected aerosolized material, even of as little as one organism, according to one report. The ease with which this microorganism can be harvested and its infectious properties make it an agent that terrorists could efficiently use.

Symptom onset begins at ten to twenty days and is usually self-limiting. Q fever disables and could cause panic but not death. Q fever usually lasts two days to two weeks and is characterized by fever, headache, fatigue, and, in some cases, pneumonia and pleuritic chest pain.

Treatment involves antibiotic therapy and supportive care. When untreated, the illness is incapaciting but does not usually cause death. In rare cases the infection may cause endocarditis, which, if untreated, is potentially fatal. Caregivers need only protect themselves from a contaminated environment. Decontamination is accomplished with soap and water or mild hypochlorite solution.

■ **SUMMARY** ■

Q fever **(Coxiella burnietii rickettsia)**

Onset of symptoms: Symptoms appear in 10–20 days.

Types of symptoms: Q fever lasts between 2 days to 2 weeks and consists of fever, headache, fatigue, pneumonia, and pleuritic pain. It rarely causes endocarditis and death.

(continued)

Types of dispersion: Aerosolized material is inhaled.
Personal protection and decontamination: Patients do not pose a risk to medical caregivers, but a contaminated environment or equipment can be decontaminated with soap and water.

Salmonellae (*Salmonella typhimurium*)

Salmonella causes one of the most common types of food poisoning. Although this bacteria has not been used as a military biological weapon, it has been used for domestic terrorism. (For example, it has been used against a group of people eating at a fast food salad bar.) Naturally occurring infections of salmonellosis are caused by ingesting the organism in food contaminated with infected feces.

Terrorist use of salmonella to harm a target group would be simple. Once a food source has been identified as contaminated, it could be easily mixed with noncontaminated food and distributed to victims. Since the bacteria is prevalent in meat products and poultry, using meat by-products to cultivate it is a reasonable way to obtain and produce enough of the microbe to cause harm to a large number of people. Spreading the bacteria on foods that are not eaten cooked is one way to ensure that it will not be destroyed prior to ingestion.

After ingestion of contaminated food, symptoms begin within eight to forty-eight hours. The victim experiences fever, headache, abdominal pain, and watery diarrhea that may contain blood and mucous. If the infection localizes, the results may be endocarditis, meningitis, pericarditis, pneumonia, or abscesses.

Infected victims can cause secondary infection if caretakers do not protect themselves from body fluids. The bacteria are killed with heat or hypochlorite solutions.

■ **SUMMARY** ■

Salmonellae (Salmonella typhimurium)
Onset of symptoms: Symptoms occur in 8–48 hours.
Types of symptoms: Symptoms include fever, headache, abdominal pain, and watery diarrhea. Localized infection may result in endocarditis, meningitis, pericarditis, pneumonia, or abscesses.
Types of dispersion: The bacteria may be dispersed by contaminating raw food or water sources.
Personal protection and decontamination: Infected patients may contaminate others. Exercise body fluid precautions and clean contaminated objects with hypochlorite solution.

■ VIRAL AGENTS

Smallpox (*Variola virus*)

Smallpox was all but eradicated by 1980, virtually eliminated through a combination of extensive vaccinations and strict quarantine. To this date, there have been no further outbreaks of small pox reported in the world. However, there are two known deposits of the variola virus (smallpox virus) remaining, one at the Centers for

Disease Control in Atlanta, Georgia, and the other at Vector in Novosibirsk, Russia. The World Health Organization has recommended that existing viruses be destroyed by 1999.

Because the disease has been eradicated does not mean that it no longer poses a threat in the United States. Other countries have experimented with the smallpox virus for biological warfare, and it is unclear whether other governments or organizations may still possess the virus for cultivation. The last vaccination given to the United States general public was in 1980; these vaccinations are no longer effective. The exposure of a segment of the U.S. population to the virus would be devastating.

The ease with which this virus is produced, and its extreme aerosolized virulence (toxicity) make it an ideal weapon. Terrorists able to secure even a small portion of the virus could, in a short time, cultivate enough to infect thousands of people. An attack of this nature would overwhelm the healthcare system and stimulate a panic that would be felt nationwide, if not worldwide.

Once exposed, a victim develops a rash, which is followed by red, raised papules that eventually form pustular vesicles. The vesicles are more common on the extremities and face, and are the opposite of those produced in chicken pox, which concentrate on the chest and abdomen. The patient remains extremely contagious for up to fourteen days after the onset of symptoms and should be isolated until the scabs separate. Caregivers must exercise complete blood, body fluid, and respiratory protection.

■ **SUMMARY** ■

Smallpox (Variola virus)

Onset of symptoms: Symptoms appear in 12 days.

Types of symptoms: Symptoms include fever, vomiting, headache, weakness, and backache. Macules, papules, and eventually pustular vesicles follow the initial rash.

Types of dispersion: The virus is airborne through aerosolization.

Personal protection and decontamination: Infected patients are contagious for up to 17 days after onset.

Venezuelan Equine Encephalitis (VEE)

VEE causes disease in horses, mules, burros, and donkeys. It is typically found in South and Central America, Mexico, Trinidad, and Florida, and occurs as a result of a bite from a mosquito carrying blood from an infected animal.

When the disease occurs naturally, there always is an outbreak among the equidae (horses, mules, etc.) population before humans are affected. If the virus is spread by unnatural means, an outbreak will not initially involve equidae during its early stages. The occurrence of the disease outside of its natural geographical area is another clue that the outbreak is not of natural origin.

VEE was prepared as a weapon by the United States during the 1950s but was later destroyed when the biological weapon program was terminated. Other countries may have experimented with VEE; the extent to which those countries have been successful is unknown. It is reasonable to expect, however, that since this virus is easy to acquire, cultivation by terrorists presents a threat.

Once a victim is exposed, an incubation period of between one and five days is followed by a sudden onset of symptoms. VEE generally establishes an infection in the meninges of the brain and within the brain itself, so symptoms will coincide with this inflammation. Initial symptoms include generalized weakness and numbness of the legs, photophobia, and severe headache. As the infection progresses, the symptoms advance to nausea, vomiting, and diarrhea. Children can have more severe central nervous system symptoms, including coma, seizures, and paralysis, leading to a 20% chance of death. Unborn fetuses are endangered by encephalitis, placental damage, spontaneous abortion, or congenital anomalies.

No specific treatment exists except for supportive care that may include analgesic medicines. If seizures result from the infection, anticonvulsant medications may be required. The most severe phase of the infection lasts between twenty-four to seventy-two hours, and may be followed by lingering or permanent paralysis or lethargy.

Blood and body fluid precautions should be exercised when caring for an infected individual. Decontamination of equipment can be accomplished using a hypochlorite solution or by destroying the virus with heat (80°C for 30 minutes).

■ SUMMARY ■

Venezuelan equine encephalitis **(VEE)**
Onset of symptoms: Symptoms appear in 1–5 days.
Types of symptoms: Symptoms include generalized weakness, numbness of the legs, photophobia, and severe headache. The most severe phase lasts 24–72 hours.
Types of dispersion: The virus is an airborne aerosol.
Personal protection and decontamination: Blood and body fluid precautions should be exercised. Decontamination is accomplished with a hypochlorite solution.

Viral Hemorrhagic Fevers (VHFs)

VHFs are a group of viruses that cause uncontrollable external and internal bleeding. Since ebola has been seen in the news many times in recent years, these types of infections have the potential to cause hysteria in a population. Terrorists would most likely trigger this hysteria through false threats and hoaxes rather than through using one of the viruses. But the possibility of obtaining possession of these viruses and then using them always exists.

VHFs are a group of illnesses caused by ribonucleic acid (RNA) viruses of several families, including Ebola and Marburg virus (Filoviridae family), Lassa fever, Argentine and Bolivian fever (Arenaviridae), Hantavirus genus, Congo-crimean hemorrhagic fever virus (Bunyaviridae family), and yellow fever virus and dengue hemorrhagic fever virus (Flaviviridae family).

These viruses, in general, cause permeability and loss of intravascular fluid in the vasculature. Early signs of infection are fever, pain, and conjunctival subcutaneous bleeding. As the infection progresses, mucous membrane hemorrhaging, pulmonary involvement, and shock result.

Depending on the extent of the infection, and the virus responsible, the mortality rate can be as high as 90%. Ebola and Lassa have the highest mortality rate and the most rapid onset of symptoms.

These viruses are transmitted in a variety of ways, but the most dangerous is by aerosolizing the agent. It is conceivable that these viruses could be used as a warfare agent, but the difficulty in obtaining and cultivating such organisms makes them an unlikely weapon for terrorist organizations.

■ SUMMARY ■

Viral hemorrhagic fevers (VHFs)

Onset of symptoms: Onset is varied depending on the specific virus.

Types of symptoms: Symptoms include fever, easy bleeding, hypotension, and shock.

Types of dispersion: Viruses are transmitted by contaminated food sources or aerosol dispersion.

Personal protection and decontamination: Complete body isolation is necessary. Decontamination is accomplished with phenolic or hypochlorite solutions.

BIOLOGICAL TOXINS

Botulinum Toxin

Botulism is a serious and occasionally fatal disease caused by a toxin produced from an anaerobic bacterium, *Clostridium botulinum*. The bacteria are found in poorly handled food and account for many cases of food poisoning (botulism) in the United States. This toxin could be easily produced and spread over a targeted area as an aerosolized particle. The effects of inhaling it are similar to those from ingesting it, except that the onset of symptoms may actually be slower.

Botulinum toxins are generally believed to be the most poisonous substances yet discovered. Toxic testing has revealed that it takes less than 0.001 μg/kg of body weight to kill 50% of a test animal population. That can be calculated to a lethal dose for a 220-pound man to be as little as 0.1 μg. As few as 950 molecules of toxin can be lethal for white mice. When these compounds are compared to other antipersonnel chemical weapons, their toxicity is placed in perspective. For example, botulinum toxin is 15,000 times more toxic than VX, the most toxic of known nerve agents.

Botulinum causes injury to the body by bonding to the nerve endings at the synaptic junction. This action prevents the nerve's terminal end from releasing acetylcholine that is needed for nerve impulse transmission. The effect is exactly opposite that of organophosphate nerve agent poisoning. During organophosphate poisoning, the synaptic junction is filled with acetylcholine because the organophosphates bind with the enzyme acetylcholinesterase, causing a continuous transmission of nerve impulses.

The inability to transmit nerve impulses to the skeletal muscles and parasympathetic system causes generalized weakness, dry mucous membranes, and eventually respiratory failure. If the concentration of toxin is great enough, as seen in ingestion injury or concentrated inhalation of aerosolized toxin, the onset of symptoms occurs as early as eight to thirty-six hours. Lower doses may cause an onset of symptoms days after exposure.

Signs and symptoms include dry mouth with difficulty speaking and swallowing; severe cases may cause the loss of gag reflex. Vision may be blurred and include double vision, photophobia, and dilated pupils. As the poisoning continues, the muscles become flaccid and develop paralysis, leading to respiratory failure and death. During the 1940s and 1950s, the mortality rate for botulism disease approached 60%, but due to improved respiratory care, the mortality rate is now below 20%.

Emergency treatment includes securing a good airway or endotracheal intubation, if necessary, and positive pressure ventilations. Hypotension may also result, so insuring that blood pressure is also maintained may be an additional consideration. Antitoxins are currently available, but risky to due to the high rate of anaphylaxis after administration. Other antitoxin therapy is under development and may be available soon. Definitive care is focused on supportive measures and can continue for weeks to months, depending on the patient and level of exposure.

■ **SUMMARY** ■

***Botulinum toxin* (Botulism)**
Onset of symptoms: Symptoms occur in 8–36 hours.
Types of symptoms: Symptoms include dry mouth, difficulty speaking and swallowing, blurred vision, dilated pupils, flaccid muscles, paralysis, and respiratory failure.
Types of dispersion: Botulism may be spread by aerosol dispersion.
Personal protection and decontamination: Protection from blood, body fluids, and respiratory exposure are necessary.

Staphylococcal Enterotoxin B (SEB)

SEB represents another form of food poisoning that incapacitates victims but rarely causes death. Clinical effects are caused from toxins produced by the bacterium *Staphylococcus aureus*. Illness caused by SEB is much more common than that caused by botulism. Natural outbreaks are usually clustered and can be traced to one exposure source, such as a restaurant or picnic. Its usefulness in combat is related to its ability to render up to 80% of exposed victims incapacitated and unable to function. Terrorists may find SEB simple to produce and easily used to contaminate open food sources and water supplies. This exposure could affect hundreds or thousands of persons, taxing local healthcare facilities and generating desired publicity. The effects of the exposure can last several weeks, causing a long-term strain on the healthcare infrastructure.

Depending on the route of exposure, the effects of SEB are different. Inhaled SEB causes systemic injury that can lead to septic shock. Ingestion of SEB causes a slower onset of symptoms that are generally less dramatic and less serious. The mechanisms of toxicity are complex and involve the use of the body's own antibody response to cause injury.

The onset of symptoms, from inhalation exposure, is from three to twelve hours and is characterized by fever, headache, muscle pain, shortness of breath, and, occasionally, chest pain. A gastrointestinal exposure may also include nausea, vomiting, and diarrhea.

Treatment is supportive in nature. Emergency responders must exercise good airway management using supplied oxygen. Severe cases may cause respiratory depression or arrest, so endotracheal intubation is recommended and aggressive positive pressure ventilations should be performed.

Decontamination of exposed equipment can be accomplished using a hypochlorite solution. Vigorous hand washing is recommended and care must be taken not to eat any food that may be contaminated with the toxin.

■ **SUMMARY** ■

Staphylococcal enterotoxin B **(SEB)**
Onset of symptoms: Symptoms appear in 3–12 hours
Types of symptoms: Symptoms include fever, headache, muscle pain, shortness of breath, and chest pain.
Types of dispersion: SEB may contaminate food and water.
Personal protection and decontamination: Vigorous hand washing is recommended. Decontaminate with hypochlorite.

Ricin

Ricin is a biological toxin formed in the seed of a castor plant (*Ricinus communis*). Castor plants are grown throughout the world and are unregulated, so this toxin is important to review. Castor beans are used in the production of castor oil, which is produced by cold crushing of the bean. Castor oil is nontoxic and used as a cathartic; however, the remaining mash contains toxic levels of ricin. There is enough ricin in the ingestion of two to four beans by an adult or one to three beans by a child to cause poisoning and death. For military or terrorist use, ricin can be produced as a solid and aerosolized to affect large numbers of victims. Once the ricin gains access to the body, the globular glycoprotein attaches to the cell wall and is eventually internalized by the cell. Ricin then blocks protein synthesis, thus killing the cell.

When inhaled, ricin causes cellular damage in the respiratory system, leading to the necrosis of tissue, bronchitis, interstitial pneumonia, and pulmonary edema. The onset is rapid, with symptoms seen in as little as three hours, but more commonly in eight to twenty-four hours. When ingested, ricin damage is seen in the gastrointestinal system and is evidenced through gastrointestinal bleeding. These symptoms are followed by cardiovascular and systemic injury, including vascular collapse, hepatic damage, renal necrosis, and death.

Emergency medical treatment is supportive and involves good airway management and fluid replacement, if necessary. There is no antidote for ricin toxicity. If an exposure to this toxin is suspected or known, proper protective measures must be taken by the rescuer, including skin and respiratory protection. Decontamination of the patient can be accomplished by washing with a hypochlorite solution.

■ **SUMMARY** ■

Ricin **(Ricinus communis)**
Onset of symptoms: Symptoms occur in 3–24 hours.
Types of symptoms: Inhalation exposure causes tissue necrosis, bronchitis, interstitial pneumonia, and pulmonary edema. Ingestion causes gastrointestinal necrosis and bleeding, followed by cardiovascular and systemic effects, including vascular collapse, hepatic damage, renal necrosis, and death.
Types of dispersion: Ricin may be dispersed through contaminated food and water sources.
Personal protection and decontamination: Secondary exposure is not a concern. Contaminated surfaces can be decontaminated with a weak hypochlorite solution.

Tricholthecene Mycotoxins (T2)

T2s are chemically complex toxins that are naturally produced by fungi. There are more than forty such mycotoxins capable of inhibiting protein synthesis and destroying the integrity of the cell membrane. The toxin targets the most rapidly reproducing cells, so those found in the skin, mucous membranes, and bone marrow are most at risk.

T2 can enter through all routes, causing injury along whichever ones are exposed. Symptoms appear within minutes and are evidenced by burning, itching, reddened skin, burning in the nose and throat, sneezing, burning of the eyes, and conjunctivitis. Burning and reddening rapidly advances to blackened necrosed tissue. Gastrointestinal symptoms include nausea, vomiting, diarrhea, and abdominal pain.

Rescuers must wear chemical protective clothing. All clothing worn by a victim must be removed to lessen exposure and prevent secondary exposure of the medical provider. This toxin is extremely hearty, requiring a heat source of 500°F for 30 minutes to destroy. The most efficient detoxifying agent is sodium hypochlorite, which can eliminate the toxic activity of T2. The patient should then be washed with soap and water to remove residual material. T2 is non-water-soluble, so complete decontamination will require careful, complete washing.

There is no specific antidote to T2 poisoning and supportive care has limited benefit. If T2 is ingested, activated charcoal should be given.

■ **SUMMARY** ■

Tricholthecene mycotoxin **(T2)**
Onset of symptoms: Symptoms appear within minutes to hours.
Types of symptoms: Symptoms include burning, itching, reddened skin, burning in the nose and throat, sneezing, burning of the eyes, and

conjunctivitis. Gastrointestinal symptoms include nausea, vomiting, diarrhea, and abdominal pain.

Types of dispersion: T2 is dispensed in aerosol form (reportedly as yellow rain).

Personal protection and decontamination: Complete skin and respiratory protection is necessary. Patient isolation is not required. Wash contaminated equipment with hypochlorite solution.

SUMMARY

If a terrorist chooses to use a biological agent to inflict harm on a population, emergency responders will not only be in great danger from the product but also at a loss to detect and determine what toxin is used. For the terrorist to complete the goal of causing fear within a targeted population, he/she will, most likely, take responsibility for the attack and advise the victims of their exposure. The only other clues will come from multiple patients complaining of similar or identical symptoms. In this case, the emergency responders' familiarity with the symptoms of each toxin may guide them in the proper response and care. Some examples of recent attacks with biological toxins include:

- In September, 1984, in northwest Oregon, a cult placed salmonella on area salad bars. Their overall goal was to make enough people sick so that a voting campaign against the group would have poor turnout.
- In 1992 in Minnesota, a local militia cultivated ricin for dispersal against law enforcement and local government.
- In May 1995, the Aum Supreme Truth targeted New York City with biologicals to be dispersed over the city by using skyscrapers as dissemination points. There have been reports that this same group is in search of the ebola virus for weapons use.

The likelihood that a biological toxin could be used on the American population is real, due to the ease of acquiring these substances. Anthrax, for example, considered by many as the Saturday Night Special of terrorists, is extremely stable and easily acquired and cultivated. Once dispersed, the toxin would create flu-type symptoms in the population. Approximately three days later, if antibiotics were not given, patients would feel better. By day four, patients would get worse and die.

It is estimated that Third World countries and the old Soviet bloc have approximately one hundred different biologicals that could be used in weaponry. The exact list is unknown; however there has been much speculation on strains and genetically altered biologicals. Also unknown is the possibility of marrying bacteria to host viruses such as the ebola virus within the anthrax bacterium. To date, we do not know whether this type of "black biology" has been accomplished.

5

NUCLEAR TERRORISM

OVERVIEW

Most studies indicate that a terrorist threat in the form of a thermonuclear bomb is highly unlikely; in fact, other forms of nuclear terrorism may be much easier to carry out. A nuclear bomb threat is unlikely to be carried out for a number of reasons, including its extreme expense, its logistical difficulty, and the enormous amount of technology necessary to develop and disperse such a device.

The true threat comes from a "dirty bomb" or explosive dispersion device, a conventional explosive wrapped or impregnated with radioactive material, the release of which is intended to contaminate an area or population. Other nuclear terrorist acts might consist of bombing attacks on storage facilities, nuclear plants, or nuclear material transportation vehicles. Because of the obstacles that must be overcome, nuclear terrorism does not appear to be a viable means of intimidation. On the other hand, Americans fear radiation, and such an anxiety reaction would make a terrorist act successful. Is all of the difficulty worth it to a would-be terrorist? Only time can answer that question.

THERMONUCLEAR DEVICES

With the recent fall of the Communist government in Russia, it would seem plausible that a splinter group might obtain a nuclear device. Along with the problems of using such a device, there is no substantiated proof that terrorists have seriously attempted to possess one. This is not to say that, given time, obstacles to developing or obtaining a nuclear device might not be overcome. Research has revealed that the threat is becoming more real every day. Black-market uranium and plutonium in small amounts have already been found and will probably become more available in the future. The technology to produce a thermonuclear device has also become easier to obtain and use. Most experts agree that it is just a matter of time before a desperate group will obtain or build a nuclear device capable of causing harm to a large group of people.

DIRTY BOMBS

Radioactive material is used in numerous ways to benefit society, from medical care to industrial soil density testing to laboratory experimentation. It is so commonplace that its transportation and storage occurs throughout the United States. The use, storage, and disposal of radioactive material is more highly regulated than the handling of hazardous materials. This is due to fear: Ionizing radiation is dangerous. It is capable of causing destruction to living cells and changing DNA structure, thereby causing harm to future generations. Terror created by the threat of radioactive contamination would be great.

Since the material is found in so many places in communities, it would be an easy task to obtain and disperse it in a populated geographic area. Radioactive material for medical use usually has a short half-life (the time it takes to reduce radioactive strength by one-half) and does not pose a serious long-term threat. Radiation used in industry, on the other hand, can be very potent and have a half-life of hundreds to thousands of years. Combining radioactive material of this nature with a dispersion device could render an area uninhabitable for generations to come. This type of terrorism would attract worldwide attention to the terrorist's cause.

BOMBINGS INVOLVING STORAGE FACILITIES OR TRANSPORTATION VEHICLES

Easier and safer for the terrorist would be a detonation at a nuclear storage facility, power plant, or transportation vehicle carrying nuclear material. Most nuclear power plants guard against the threat of bombings. Nevertheless, the potential still exists for an attack that could cause a release of radioactive material. Most experts believe that this type of threat is minor and, if carried out, would most likely be unsuccessful.

Bombings involving storage facilities or transportation vehicles pose the greatest threat. High-level radioactive material is normally stored in well-protected and shielded storage areas, but transportation to and from these areas leaves the material vulnerable to attack. In 1998, the removal of spent radioactive fuel from nuclear power plants throughout the nation is scheduled to begin. The spent fuel will be transported from the power plants to a burial ground in the western United States. During the transportation, the fuel will travel through many states and communities, creating a heightened risk for either an accident or terrorist attack. Moving spent radioactive fuel across the country provides one of the largest targets for nuclear-based terrorism.

TYPES OF INJURIES TO EXPECT FROM EXPOSURE

Victims of radiation exposure rarely show immediate signs or symptoms. Emergency responders must evaluate the scene and look for any evidence of radiation exposure. Radiation injuries are caused by external irradiation, contamination, incorporation, or a combination of the three.

External irradiation is the injury that occurs when radiation (gamma or x-ray) passes through the body from an external source. The intensity of the injury depends on the amount of exposure to the source. Once the victim is externally exposed, he/she does not become radioactive and can be handled by the rescuers without fear of receiving a radiation injury.

When a patient becomes partially or wholly covered with radioactive material, it is referred to as a contamination injury. The contamination can be in the form of a solid, liquid, or gas, and can be either external or can enter the body through the skin, lungs, open wounds, or the digestive tract. Injury caused by contamination will be internal, external, or both. Rescuers handling the contaminated patient can themselves become contaminated, and need to remember the rules of time, distance, and shielding explained in the next section. If external contamination exists, decontamination of the patient must be done prior to transportation to the hospital.

Incorporation is the uptake of radioactive material by tissues, cells, and organs. Based on its chemical properties, the material is attracted to areas of the body where it can be easily incorporated; for example, radium seeks out bone and iodine seeks out the thyroid. It is this principle that allows medicinal radioactivity to be used to diagnose and treat these target areas. In fact the antidote for exposure to radioactive iodine, potassium iodide, blocks the thyroid gland's uptake of radioactive iodine and would thus help reduce thyroid disease caused by exposure to the substance during a nuclear accident. Incorporation cannot occur unless contamination has occurred.

Gamma irradiation of a part or even the whole body does not usually require immediate emergency treatment. Signs and symptoms usually appear days, weeks, or months after the injury takes place.

The type of radiation, how much of the body was exposed, and the dosage determine how much radiation sickness a victim will suffer. Depending on the dosage, the effects of radiation on the body are varied. The skin may swell, blister, redden, flake, or itch. Breathing may be affected because of swelling or damage to the alveoli. Other injuries include permanent or temporary sterility, loss of menstruation, reduction of sperm count, damage to blood vessels, cancer, and genetic changes.

Recovery from large doses of radiation may require months or years. Recurrent or chronic problems such as chromosomal damage and reproductive difficulty may last a lifetime and affect future generations.

PERSONAL PROTECTION

The basic rules of protection for both emergency responders and victims are easily understood but sometimes difficult to carry out. These include time, distance, and shielding. Adherence to these rules will greatly reduce exposure.

Time

The time factor is simple: *Limit your time of exposure.* The longer the exposure to a source of radiation, the greater the damage to the body tissues. If rescue or extrication attempts require that an extended amount of time be spent in an area affected by radiation, then personnel should be rotated throughout the area, limiting individual exposure.

Distance

Distance, the second protection rule, is also simple: *Stay as far away as possible.* Doubling the distance from the source decreases exposure by a factor of four. For example, an exposure of 16 milliroentgen per hour at a distance of 3 feet would be reduced to 4 milliroentgen per hour at a distance of 6 feet. But the opposite also holds true: An exposure of 8 milliroentgen per hour at a distance of 4 feet would be 32 milliroentgen per hour at 2 feet. Therefore, it is important to realize that exposure to radiation is always reduced by moving away from the source.

Shielding

Shielding is the third rule of protection. *Keep something between you and the source.* Shielding is based on the fact that the denser the material, the more radiation it blocks. Lead is the most common shield used around X-ray equipment because of its high density. At the site of a radiation emergency an automobile or a hill of dirt can be used. Heavy clothing will stop all alpha particles and some beta particles. Lead shielding, although not practical for rescue, is effective against gamma rays and X-rays.

Monitoring

If use of a radioactive material is suspected during a terrorist attack, monitoring must be part of the initial survey. Radioactivity has no odor or taste and cannot be seen, so a responder will not normally know when danger is present. It is only by approaching the scene with a high degree of suspicion, and appropriate monitoring equipment, that a responder will be prepared for hidden danger. Most hazardous materials response teams carry some instrumentation to detect dangerous radiation. If one of these teams is not available, other immediate sources for instrumentation may include nuclear medicine departments at hospitals, or laboratories that use radioactive isotopes in their experiments.

SUMMARY

Although nuclear terrorism seems to be a viable threat, the fact is that, in order for such an attack to succeed, enormous technological barriers would have to be overcome. A terrorist nuclear threat will always be possible; its probability, however, is slight. Admittedly, terrorists are not afraid of using extreme acts of violence. Where we are truly vulnerable, though, is in our nuclear reactors and in our medical and industrial laboratories. However, there has never been a nuclear bomb threat from a terrorist group. If we think of the Gulf War as a milestone in the history of weapons of mass destruction, we see that while Iraq had some of the technology necessary, their devices were similar to those tested during the World War II era, weaponry that would require more fissile material and an enormous quantity of energy to initiate a reaction. Current thinking about the possibility of terrorist nuclear weapons is based on the fact that such devices need uranium 235 in sufficient quantities and of superb quality; this material is unavailable on the black market. In addition to the material required for such an undertaking, the variety of disciplines required to produce a nuclear device would be prohibitive.

Realistically, common radiological isotopes found in industry and medicine would be the motivated terrorist's choice. Even so, the restrictions on these substances are rigorous, and they are not usually available. This does not mean that such a threat will never exist; dirty bombs using common munitions for the dispersal of radioactive isotopes are real possibilities. An example occurred in Houston, Texas, on February 26, 1996: Two cobalt-60 cameras and a iridium-192 camera were stolen. The Ir-192 camera had decayed to minimal levels of radioactivity while the Co-60 cameras contained levels ranging from 35.6–8.6 curies. The equipment was eventually recovered, but not before the incident had great emotional impact on the community, the fire department, the police department, and a variety of community support agencies. Although this radioactive material was not stolen for the purpose of terrorism, it demonstrates how potentially dangerous radioactive material can be obtained for terrorist use. In this case, the material was recovered without an injury occurring.

The United States has trained specialized units to handle nuclear emergencies. The Nuclear Emergency Search Teams (NEST) of the Department of Energy (DOE) have been established to identify and measure radiological sources. Their tasks, in conjunction with the FBI and FEMA include, but are not limited to, searching for radiological sources, defusing nuclear weapons, decontaminating victims and property, and repairing collateral damage. These experts are invaluable resources in the event of a nuclear threat.

6

PERSONNEL PROTECTIVE EQUIPMENT FOR TERRORISM

OVERVIEW

The emergency responder must consider what equipment is immediately available for personnel protection against a terrorist attack. Since the nature of this type of attack is to surprise and cause fear in a population, it will probably come when it is least expected. It is the unexpected nature of this kind of event that mandates the need for an emergency responder to utilize personnel protection immediately on arriving at a scene.

A terrorist event is not a typical emergency. It is intended to cause harm to people. Often the terrorists include emergency responders in their equation for harm. In fact, recent statistics indicate that these attacks are not aimed at one group, but increasingly include the emergency responder as part of the target. What better way would there be to destroy the confidence citizens have in their government than to tamper with the help they have come to expect when they dial 911.

Fire personnel are generally equipped to protect themselves from such disasters as fires, spills and leaks of hazardous materials, and building collapses. But they are usually not prepared for chemical, biological, nuclear, or explosive devices directed at them. In most emergency response cases, firefighters have time to assess a situation. Even the emergencies that require a rapid response, such as a house fire or auto extrication, allow time for planning. In the case of a terrorist event, the emergency responder may become part of the incident before any plan can be enacted. Only a well-trained and prepared emergency responder can render help instead of becoming a victim.

What personnel protection will safeguard the emergency responder? In the case of a typical emergency medical services (EMS) unit such as an ambulance crew, not much protection is available. These units are not generally equipped with personnel protective equipment beyond what is needed for safe handling of blood and body fluids. These emergency responders may best protect themselves by recognizing danger and retreating to a safe location.

Fire department units take personnel protection a step further. Not only do they have carry turnout gear that can provide some protection from nuclear, biological, and chemical weapons, but they also carry protection from airborne hazards, self-contained breathing apparatus (SCBA). Although this equipment was not designed to provide protection against these weapons, it is usually sufficient for use in a quick retreat and even in a quick rescue.

The next level of response comes from hazardous materials (HazMat) teams. HazMat response units carry partially or fully encapsulating suits that isolate the wearer from the environment, providing protection from most dangers found in the outside atmosphere. It is this type of protective equipment that may be the most useful for civilian emergency responders.

So what is appropriate protection against the lingering hazards of a nuclear, biological, or chemical attack? In a perfect world, the means of protection would be selected from a variety of available options, each designed for a specific type of hazard. In fact, this is the case with HazMat response, where a chemical involved in an incident is researched and a suit selected to provide the best protection against it. But this may not be possible if a terrorist released a weapon with unknown properties. The best a responder can do is to know what equipment is immediately available to provide some level of protection during and after an attack.

Chemical, biological, and, to some degree, radioactive materials enter the body by absorption through the skin, eyes, or respiratory system. The properties of the chemical determine how rapidly it gains access. This principle also holds true for biological toxins or organisms. Generally speaking, chemical warfare agents are most efficient when they are inhaled. This is an important point because virtually every fire department stocks a supply of SCBAs. These breathing apparatuses are extremely efficient in providing protection from inhaled chemical and biological agents, and have several advantages, including positive mask pressure, 21% oxygen even in oxygen-deficient atmospheres, and no mobility-limiting umbilical. On the other hand, these units have a limited air supply and are cumbersome in confined areas. For the quick rescue of victims and for personal escape, they provide excellent protection from respiratory exposure.

FIREFIGHTER PROTECTION

Firefighter turnout gear is not rated against chemical, biological, or nuclear radiation. Nonetheless, it does provide some protection against these agents, which can gain entrance to the body by contacting exposed skin. Turnout gear leaves natural openings, exposing skin to toxins in the environment. Exposure of such areas as the neck and back of the head may cause immediate injury. If the skin is mostly covered, the chemical must be in a gaseous or aerosolized form to move through the natural openings and contact the skin. Absorption of agents in gaseous form is limited, however, because less concentrated amounts reach the skin on the inside of the gear. Thickened agents must be in a strong enough concentration to penetrate the fabric of the gear and reach the skin. Penetration through the natural openings at the neck, wrists, waist, ankles, or zipper openings is also a possibility.

In liquid form, the greatest amount of the agent would settle on the outside of the gear or gain access through the natural openings. In either case, it could take time before the agent could penetrate to the inside and contact the skin. Although not considered adequate protection for prolonged exposure to a hazardous emergency scene, structural firefighting gear and SCBA offer enough protection to effect a rescue or escape. Remember, leather gloves and all of the turnout ensemble will absorb and hold chemicals and biological agents, so secondary contamination is a serious hazard. This is particularly true if the turnout is not worn properly.

RESPONSE TO A KNOWN EVENT

If emergency responders are dispatched to an event involving known chemical or biological toxins, they should make every effort to provide full protection from these agents. This will require following the procedures commonly used at hazardous materials events, such as identifying hot, warm, and cold zones; establishing decontamination corridors; and wearing the proper level of protective gear. The personnel best prepared to enter the hot zone will be those responders trained as hazardous materials technicians; other responders may provide support.

HAZMAT TEAM PROTECTION

Hazardous materials teams typically carry several levels of protection intended for use during entry into an environment containing a hazardous material. In this section, Level B and Level A gear are examined for use during a terrorist incident when a chemical or biological agent is the weapon. In either case, the handling of the incident should be very similar to an accidental hazardous materials release, except care must be taken to insure the preservation of evidence.

Level B

HazMat teams are able to provide body protection several steps beyond that of firefighters. Level B protection consists of a body suit made of saranax or similar chemically resistant material that may not be fully encapsulating. It usually does not include booties or complete head and neck coverage, and uses a secondary glove system that can be sealed with tape. Level B protects the wearer from splashing chemicals and provides respiratory protection via a SCBA. However, this garb may not completely protect the skin, leaving gaps in the neck area or in the seams that could allow a chemical to enter. Some Level B suits are very similar to full Level A suits, but are not pressure-tested to insure that all of the seams, zippers, and other openings are fully sealed. If the chemical or biological agent is not known, this level of protection may be adequate only for those responders acting as part of the decontamination team.

Level A

Level A protection consists of a totally encapsulating suit containing air. To insure that this gear is completely intact, a pressure test is done either by the manufacturer or by periodically using test equipment to locate leaking seams and punctures. Encapsulation offers excellent protection against chemical and biological agents, yet chemicals can still make entry through penetration, degradation, or permeation if the suit is not rated for the properties of the agent it is exposed to. A brief description of each of these means of entry follows.

Penetration

Penetration is the physical movement of a chemical through the natural openings of the suit. Zippers, exhalation ports, seals around the face shield, and connections can provide a route for chemicals to enter the suit. Abrasions, punctures, or tears can contribute to penetration. Most commonly, liquids, finely divided particles (aerosols), and gases under pressure can penetrate the suit.

Degradation

Degradation, the physical destruction of the suit, can be caused by temperature, concurrent chemical exposure, inappropriate storage environment, and incompatibility with a chemical. Signs of degradation are discoloration, bubbling, chaffing, shrinkage, or any visible signs of destruction.

Permeation

Permeation is the movement of the chemical through the protective materials at the molecular level. An atom has space between its center (nucleus) and its electrons (outer region); there is also space between the bonded atoms in a molecule. Chemicals can move through or permeate the protective material between these spaces. When concentrations on a suit build, either through prolonged exposure, repetitive exposure, or incompatibilities with the chemical, the chemical diffuses through the outside of the fabric. Eventually, the chemical breaks through the fabric (called desorption), leading to exposure.

MILITARY PROTECTION

Since the military has been facing this problem for decades, we can learn from their experience. Today's military chemical protective suit, called Mission-Oriented Protective Posture, Level 4 (MOPP 4) consists of boots, jacket, pants, gloves, headgear, and an APR cartridge built into a full face mask. This APR combination is called the M17. It is specifically designed to protect the wearer from chemical permeation. However, chemical penetration can occur through the natural openings of the suit. The MOPP 4 suit is lined with an absorptive charcoal layer that traps chemicals, providing several hours' to several days' protection then it is replaced with another.

The APR also absorbs chemicals into a charcoal filter, providing the wearer with clean filtered air. Again, this type of filtration is engineered to last for several days before being discarded and replaced.

Although this gear allows the wearer reasonable dexterity and mobility on the battlefield, since it is surprisingly lightweight and flexible, it has some serious drawbacks. MOPP 4 allows the wearer to drink from a canteen, a feature that is extremely important for hydration purposes, but which creates a dangerous entry point for chemicals. Also, the suit does not allow good air circulation and confines body heat, which proved detrimental during Desert Storm. Other drawbacks to the MOPP 4 gear include the possibility of chemical entry through zippers and at the waist, neck, wrists, and ankles. A rating far below 100% effectiveness makes it inappropriate for civilian use.

The increased threat of terrorism in our country leaves emergency responders looking for alternatives to typical protective gear. Some believe that military protective equipment is a viable option, but work standards and OSHA regulations do not approve of equipment with less than 100% effectiveness. On one side of the issue is the need for protective equipment that allows for good mobility and dexterity and that can be worn for long periods of time. MOPP 4 gear has been battle-proven to provide these features.The other side of this issue is strict civilian regulation enforced by law that requires 100% effectiveness of protective gear. But civilian gear, such as Level A HazMat gear, severely limits mobility and dexterity and has a small air supply, which limits work periods to 30 minutes or less. The future may bring a compromise that gives emergency responders the best of both worlds, but until that time this issue will be debated. Table 6.1 lists advantages and drawbacks of several kinds of personal protective equipment and Table 6.2 displays MOPP rated gear.

The military has long set guidelines for acceptable war casualities. In other words, a certain number of deaths are determined to be acceptable during a particular military action. This practice is unheard of for civilian responders. Can you imagine responding to a hotel fire and the incident commander determining that the acceptable life loss for the incident will be set at 30%, including firefighters and civilians? Civilian emergency responders do not accept loss of life as do their military counterparts. Each death on an emergency scene is considered a battle lost.

For this reason, military gear will probably not gain widespread use during civilian incidents. Attempts to have the military-style masks approved through OSHA have failed because of the canteen openings. Military protective equipment needs to be modified only slightly to provide the protection expected by civilian emergency responders. In the meantime, typical HazMat protective gear will be used during these events.

Table 6.1 Personal Protective Equipment

Level	*Advantages*	*Disadvantages*
Turnout gear with SCBA	Readily available; moderate level of protection	A high degree of absorption may occur. True limitations are not known.
Level B with SCBA	High degree of protection; easily placed on existing apparatus	This equipment has limited availability and has openings if not properly worn.
Level B with APR	Moderate level of protection; long-term use in hot or warm zone	You must know the exact chemical you are dealing with before you can wear an APR.

(continued)

Level A (fully encapsulating)	Highest level of protection	This equipment has limited air supply and limited mobility, and heat exhaustion occurs rapidly.

Table 6.2 Mission-Oriented Protective Posture (MOPP)

Level	*Mask*	*Overgarment*	*Overboots*	*Gloves*
MOPP-0	Carried	Readily available	Readily available	Readily available
MOPP-1	Worn	Carried	Carried	Carried
MOPP-2	Worn	Worn	Carried	Carried
MOPP-3	Worn	Worn	Worn	Carried
MOPP-4	Worn	Worn	Worn	Worn

SUMMARY

Although MOPP-4 is an excellent idea, the reality is that this gear may not be adequate as the primary means of protection against a terrorist strike in the civilian world. As concern increases, emergency protection systems are being designed and developed. The problem is to provide high-level protection to the rescuer while maintaining zero acceptable loss (meeting the OSHA standard). Unfortunately, this will necessitate the purchase of additional equipment for all emergency responders, further burdening an ever-decreasing budget.

Until this occurs, first responders must be aware of location, training, and limitations for high-level PPE. This may, in fact, increase the HazMat capability within all communities, enhancing existing services, and making HazMat ensembles available to each employee like turnout gear.

Limitations still exist. Heat stress and lack of mobility will decrease effectiveness of the wearer. These issues are vitally important to address during preplanning. Also, instituting strict personnel rotation schedules will create discipline problems at the scene and will necessitate a large cadre of personnel to handle a terrorist event.

Experts state that Level B with the appropriate APR will provide a high degree of protection most of the time under nuclear, biological, and chemical (NBC) weaponry conditions. This may be true; however, more research and development studies of personnel protective systems must take place before a definitive answer can be given. Until then, err on the side of safety and prepare using existing high-level PPE, thorough decontamination procedures, and preplanning across all emergency response disciplines.

7

DETECTION DEVICES

OVERVIEW

Contaminants at the scene of a terrorist or hazardous materials incident can present a significant danger to response personnel, since the detection of weapons of mass destruction is not an easy task. This is especially true for agents such as military chemical and biological weaponry. Identifying and qualifying possible contaminants is complicated by the natural variety of chemicals and biologicals that occur within the environment. Additional problems are created by the chemical detection systems currently used by hazardous materials teams, which are incapable of extremely low-level detection. These problems are further complicated by the fact that antipersonnel chemical weapons are very toxic at levels well below current detection system capabilities. Biological weapons, for example, cannot presently be detected using commercially available detection equipment.

New technology breakthroughs in biological detection are encouraging for the future of detectors. Until these detectors are available, emergency responders must rely on environmental clues and signs and symptoms presented by victims. For example, an incident that occurs around a public assembly, industrial process, controversial business, or on a significant date (historical event) should be considered suspicious. Furthermore, events where there are multiple requests for assistance; multiple patients with similar signs and symptoms; and dead birds, animals, or insects should alert emergency responders to the possible presence of chemical or biological agents. Unfortunately, this is not a good method for detecting chemical and biological agents. Some agents, such as nerve toxins, produce symptoms almost immediately, while others, like mustard gas, are slower acting. When immediate symptoms are not apparent at the scene of an exposure, it may be difficult to determine whether a terrorist event has taken place.

Although detection devices currently carried by hazardous materials responders may not perform to the level needed to positively detect warfare chemicals, they can be used to provide clues to the event. The following is an overview of monitoring equipment commonly carried on hazardous materials units, as well as military detection devices available for use.

DETECTION DEVICES

Electrochemical Devices

Electrochemical sensors have two electrodes, one to sense and the other to count. The sensing electrode maintains a constant electrical potential, while the counting electrode measures the electromagnetic potential. This potential is directly proportional to the partial pressure of the chemical the sensor is designed to detect. The sample is introduced through a semipermeable membrane to the solution within the sensing electrodes. The electrodes, in association with the chemical present, set up an electrical gradient, which is then measured by the instrument and, through an amplifier, is displayed as a percentage of the chemical.

Some instruments on the market are combination devices. In these instruments, there are several electrochemical sensors, one for each chemical that the instrument can detect. In each case, the electrochemical sensor creates an electrical current that is proportional to the contaminant. These instruments can monitor two or more gases, but like the single-sensor monitor, they do not identify the atmosphere. All the monitor will do is alert the user to the presence or absence of the chemical that the monitor is set up for.

Ozone, chlorine, hydrogen sulfide, fluorine, carbon dioxide, and bromine can all neutralize the alkaline electrode sensor bath. When this occurs, a false positive is observed. When the alkaline bath has been neutralized, the sensitivity of the instrument has also been reduced. The limitations of these instruments are sensitivity to atmospheric changes, neutralization of the alkaline electrode sensor yielding false positives, and the interference of gases that may also give false positives. Detection limitations are determined by the way the instruments are manufactured and the chemical sensors they contain.

Combustible Gas Indicators (CGIs)

CGIs use a Wheatstone bridge circuit as their sensor. Within the circuit are two filaments; one filament is called the sensor and the other the compensating filament. When the unit is in the on position, the filaments heat up, and the gas being tested is pumped over them. As the gas passes the sensor, the filament heats up. The compensating filament is a control for the system, which will adjust to conditions such as temperature and humidity. The difference in resistance between the sensor and the compensating filament is translated by the circuit to a meter readout.

Limitations of CGIs include the following:

1. Oxidizing gases will heat up the filament faster, giving a false high reading.
2. Filaments are sensitive to contaminants such as heavy metals, organic lead vapors, sulfur compounds, and silicone, which may cause corrosion.
3. High voltage will cause needle fluctuations or zero readings.
4. Response curve factors must be used.
5. High humidity will mask the sensor, giving slower response or a false reading.

Microprocessing Monitors

These monitors are activated by a mixture of metal oxides, temperatures, and voltage across a semiconductor sensor. Usually, they are set up to monitor a select group of gases; they cannot identify a gas, only indicate its presence. Limitations of microprocessing monitors include interference gases; high humidity; immediate changes in temperature that may cause condensation; dust; and particulate matter. All of these factors can prevent accurate readings.

Photo-Ionization Devices (PIDs)

PIDs detect concentrations of gases and vapors with an ultraviolet light that bombards contaminants, ionizing them. Matter is composed of atoms which have a cluster of electrons rotating about a central nucleus. The electrons are negatively charged. Ultraviolet light stimulates the electrons, pulling the outermost electrons away from their original path. The energy released when an electron returns to its path is termed *ionization potential* (IP). All materials have IP, which is measured by PIDs in electron volts (eV).

To detect a chemical, one must know its IP, so that the proper ultraviolet lamp may be used, one having a higher ionization potential than the contaminant. Currently, all that is known about the IPs of chemical agents is that they are under 10.6 eV. This means that a lamp with higher eV must be used, such as argon (11.7 eV) or lithium fluoride (11.8 eV). Each chemical also has a relative response pattern to the calibrate gas; relative response patterns are presently unknown for specific military chemical agents.

Humidity, interference gases, and particulate matter can all limit the detection capabilities of the PID, giving false or inaccurate readings. Other limitations are condensation, high temperatures, and maintenance cleaning. Overall concentration limitations of PIDs are 0.1–2,000 ppm.

Flame Ionization Devices (FIDs)

The FID is a low-concentration detection device, its typical detection range being 0.2–1,000 ppm, with an IP of 15 eV or less. Its principle of operation is similar to that of the PID; however, instead of light a hydrogen flame is used to ionize molecules. Again, relative response patterns must be known in order to detect the presence of a particular chemical.

Since FIDs use a hydrogen flame, oxygen must be present within the atmosphere. Other limitations include:

1. Its sensitivity to aromatic compounds is extremely low.
2. It cannot be used for inorganic compounds like sulfides or for inorganic gases.
3. Certain functional groups attached to the contaminant molecule can reduce the sensitivity of the instrument.
4. Highly skilled individuals are required to operate the FID and interpret its data.

Colorimetric Tubes

Colorimetric tubes associated with military detection devices are the detection instruments of choice. However, as has previously been noted for other detection devices, colorimetric tubes have low-level detection limitations.

A colorimetric tube is a hermetically sealed glass tube with reagents inside, which react chemically with a contaminant in a select volume of air. A predetermined volume of air is introduced into the tube, particulates are filtered out, and a known contaminant reacts with the reagent. This reaction creates a stain in the reactants that can be measured in ppm or volume.

Colorimetric tube detection limitations include the following:

1. If a stain is not generated by the reagents, it does not mean there is no contaminant. The lack of a stain may mean that a chemical was not present in a quantity within the range of measurement.
2. Detector tubes may have an error factor as great as 50%. Temperature, humidity, pressure, storage conditions, shelf life, and interfering gases can all play a role in tube accuracy.

Table 7.1 identifies which tubes may be used to detect military agents.

Table 7.1 Colorimetric Chemical Detection Tubes

Agent	*Tube*	*Detection Range*
Tabun	Phosphoric acid esters	0.05 ppm
Sarin	Phosphoric acid esters	0.05 ppm
Soman	Phosphoric acid esters	0.05 ppm
Chloropicrin	Carbon tetrachloride	1–15 ppm
Lewisite	Organic arsenic compounds	3 mg/m^3
	Arsine	0.1 ppm
Mustard	Thioether	1 mg/m^3
Nitrogen mustard (HN-1)	Organic basic nitrogen compounds	1 mg/m^3
Nitrogen mustard (HN-2)	Organic basic nitrogen compounds	1 mg/m^3
Nitrogen mustard (HN-3)	Organic basic nitrogen compounds	1 mg/m^3
Hydrogen cyanide	Hydrogen cyanide	2–30 ppm
	Hydrocyanic acid	2–30 ppm
Chlorine	Chlorine	2–30 ppm
Phosgene	Phosgene	0.02–1 ppm
Arsine	Arsine	0.05–3 ppm
Cyanogen chloride	Cyanogen chloride	0.25–5 ppm
Pepper spray	Olefins (not recommended)	1–55 mg/L
Mace	Chloroformates	2–10 ppm
	Acid aldehdyes (not recommended)	100–1000 ppm

(Courtesy of Drager Corporation, provided through their civil defense branch)

pH Papers

pH papers detect the representative acidity or alkalinity of a solution or gas. pH or positive hydronium ions are measured on a scale from 0–14, 0–6.9 being acidic and 7.1–14 alkaline. The number 7 is neutral. A chemical in contact with a pH paper changes the color of the detection paper, identifying its degree of acidity or alkalinity. pH detection papers do not give specific concentrations of material; however, degrees of color may identify general concentration levels.

Most military chemical agents are neutralized to 7. This enables a long-term shelf life for selected munitions. Unfortunately, a terrorist does not pay attention to long-term shelf life. Chemical munitions made clandestinely may be acidic. The acidity or alkalinity of a chemical will aid in a decision on decontamination solutions.

Radiation Detection

Radiation is a broad term that identifies energy transmission. It is the spontaneous decay of radioactive materials that emit particles from the nucleus of an atom. Ionized radiation produces a particle or wave energy that disrupts the atom, which bombards the matter around it and is the cause of radiation damage to cells, tissue, and organs.

There are three forms of radiation: alpha, beta, and gamma. The first two are particles. Alpha particles are positively charged and made up of two neutrons and two protons with an approximate travel distance from a source of 4 inches. Beta particles can be positively or negatively charged and are approximately the size of an electron. Travel from a source can be up to 30 feet. Gamma radiation is an electromechanical wave of energy that can travel great distances and penetrate materials, depending on their source.

Alpha-beta detectors, beta detectors, and gamma detectors can detect radiation. Geiger-Mueller (GM) detectors can identify small amounts of beta and gamma radiation. Film badges can sense beta, gamma, X-ray, and neutron radiation, while pocket dosimeters are beneficial in identifying alpha, beta, and gamma radiation.

It is suggested that scintillation and luminescent devices be used in conjunction with radiation monitors because these light-emitting devices are highly sensitive. Multiple meter detection is necessary for a terrorist attack, since the radiation level would be unknown. Even though the chance of a thermonuclear device being used is remote, it is possible. The most likely scenario is a radioactive source exploded with conventional munitions dispersing radiation into the immediate environment. Depending on the source, high- to low-level radiation is a possibility. Detection must incorporate these possibilities.

■ MILITARY DETECTION DEVICES

There are several military detection devices currently employed to identify chemical warfare agents. Unfortunately, these devices, much like the ones previously discussed, require the user to enter the hazardous environment. This limitation places the initial responder at great risk. In some cases, by the time these detection devices identify the presence of a hazardous material, the responder is in a contaminated environment.

M-8 Paper

The M-8 paper detection device is similar to pH paper. It detects the presence of a particular chemical, but does not identify its concentration level. M-8 paper is used on liquid contaminants and will detect nerve, blister, and persistent nerve and blister agents. The operator must place the paper into the liquid for approximately 30 seconds, after which time the detection paper will change color, identifying the possible agents. A yellow color indicates nonpersistent G agents, red indicates blister agents, and olive green or black indicates a persistent V agent. If insecticides, ethylene glycol, or petroleum products are present, however, the paper will indicate a false positive.

M-9 Paper

M-9 detection tape is whitish-cream in color. Once a liquid comes in contact with it, the reactant chemicals within the tape change to red, reddish brown, purple, pink, or brown, depending on agents present. The tape has an adhesive backing that adheres to equipment or protective clothing. Protective clothing must be worn when using it, since there are potential carcinogens in the indicator dye. Again, the presence of insecticides, herbicides, ethylene glycol, or petroleum products will cause false positives.

M256A1 Detection

The M256A1 detection system has six tubes and twelve detector packets, each containing enzymes that act as reagents in the presence of liquid or gaseous chemical agents. This kit can identify extremely low concentrations of chemicals. However, its use requires the responder to remain within the hazardous atmosphere for approximately 20–30 minutes. Detector packets and reaction tubes should be shielded from sunlight or heat sources, so that the reaction process will not be interrupted by evaporation of reagents. Agents that can be detected are hydrogen cyanide, cyanogen chloride, sulfur mustard, nitrogen mustard, distilled mustard, phosgene oxime, Lewisite, and V and G series nerve agents. The following list describes what each tube detects:

- Tube one (T-400). If there is no reaction, there are no blister, blood, or nerve agents present.
- Tube two (T-401). If a reaction is present, both G and V nerve agents series are possibly present.
- Tube three (T-402). If a reaction is detected, sulfur mustard is possibly present.
- Tube four (T-403). If a reaction occurs, it indicates the possible presence of phosgene oxime.
- Tube five (T-404). If a reaction occurs, it indicates the possible presence of hydrogen cyanide or cyanogen chloride. If an extremely fast reaction occurs, the presence of hydrogen cyanide or cyanogen chloride in high concentrations is more than likely.
- Tube six (T-404A). If a slow reaction occurs, extremely low concentrations of hydrogen cyanide or cyanogen chloride are present. However, if a faster reaction is present, higher concentrations of hydrogen cyanide or cyanogen chloride are indicated.

Chemical Agent Monitor (CAM)

The CAM uses a microprocessing chip to identify the presence of nerve or blister agents. Its level of sensitivity is unknown, and, in some models, a switch between nerve and blister agents must be thrown in order to read the instrument. A newer model, the field chemical agent monitor (FCAM) has a built-in processor that switches automatically between nerve and blister agents. The improved chemical agent monitor (ICAM) is a hand-held device that can detect vapors by sensing molecular ion movement and specific chemical molecular mobility or molecular velocities. The ICAM can identify type of nerve agent (both G and V series) and blister agent (HD, HN1, HN2, HN3) and differentiate between them.

Detection Devices	*Uses*	*Limitations*
PID	Initial reconnaissance	Concentration limitations; corrosive gases; humidity; condensation; high temperature; dust; relative response curves not available
FID	Initial reconnaissance	Sensitivity to aromatics; cannot be used for inorganics; certain functional groups; relative response curves not available
Colorimetric tubes	Initial reconnaissance	Detection level of contaminants limited
pH paper	Initial reconnaissance	Detects acidity and alkalinity, not concentration
Radiation detector	Initial reconnaissance	Dependant on level of detection of instrument; should have available for response; high- and low-level and neutron radiation detection
M-8	Initial reconnaissance	Liquid only, no vapor; false positives
M-9	Level of contamination on protective equipment	Liquid only, no vapor; false positives
M256A1	Initial reconnaissance	Time element for test; detection of nerve, blood, and blister agents only
CAM	Detection in decon and any contaminated areas	False positives from organic solvents
ICAM	Warning device	Gross vapor detection

SUMMARY

There are a variety of monitors available to detect chemical agents and nuclear particles; each has limitations. Military monitors are becoming available for use by MMST, but these instruments are extremely expensive. Gas chromatographs and mass spectrometers are excellent instruments for analysis. Unfortunately, these are too expensive and require a high level of training for use. However, universities and research facilities use these machines on a daily basis. Preplanning with university officials for their use, and understanding their limitations and sensitivity could be beneficial for future emergencies.

Unfortunately, there are no detection devices available for biologicals or biological toxins. The natural variety of biological materials we have within our environment, although not normally toxic, could result in false positives if a biological detection system were available. Most research in the detection of biological agents is directed toward immunoassays, biological recognition assessments coupled with microprocessing capabilities, and laser-based detection systems. However, this type of technology is in its infancy and will probably not be available to emergency responders for a few years to come. (Military projections predict availability for military use by the end of 1998; this may be an optimistic time frame.) Presently, health departments analyze health trends in populations. In many areas of the country, for example, chickens are used as indicators of possible diseases that might infect the local population. Fearing a second attack using a chemical or biological agent, Tokyo officials used caged birds for early detection after the sarin attack in the Tokyo subway. Caged birds were once similarly used in coal and gold mines during the industrial age.

Realistically, what should occur is a daily monitoring of public health, identifying current levels and health patterns as they relate to organophosphates, visicants, pulmonary agents, cyanides, lacrimator agents, riot control agents, radiation, biologicals, viruses, and biological toxins. By watching trends in international terrorist events and being alert to changes in local health status, we may have warning of exposure to clandestinely dispersed biologicals and chemicals.

■ 8 ■

DECONTAMINATION

OVERVIEW

Whether an event is determined to be a hazardous materials accident or an intentional chemical/biological event, decontamination to limit casualties will be the single most important aspect of the operation. Decontamination is the process by which chemical and/or biological agents are removed from individuals, equipment, and the environment. It is a systematic and deliberate course of action taken for a safe environment. Whether decontamination involves a single victim or a group, time is working against the first responder. Quick, decisive action is more important than the exact method by which initial decontamination occurs.

Decontamination methods can be grouped into two separate categories, each requiring a degree of deliberate action. The first involves the physical removal of the contaminant. This may or may not be followed by the second, chemical removal. In both, splash, debris, and runoff should be considered and planned for. If runoff is not captured or at least guided away from the decontamination area, workers may potentially be contaminated at the scene. Close monitoring of the runoff materials from the decontamination process will therefore ensure a safe working environment. Due to the extreme toxicity of chemical and biological warfare agents, detection and monitoring should continue even after decontamination to ensure that the process has been efficient and to stop possible cross-contamination.

TYPES OF CONTAMINATION

Type of contamination determines what decontamination steps should be taken. Chemical, biological, or nuclear contamination can occur. In each of these categories, an agent may be in a liquid, solid, gas, or a combination of these states. For example, nerve chemical agents, commonly called nerve gas, are typically dispersed as an atomized form of liquid. This liquid eventually vaporizes so that respiratory exposure can take place. Without the atomizing effect, the vaporization process occurs slowly, lessening the desired effect. Solids are dispersed in a similar manner. They are made to create a fume or fine dust so that victims can breath the toxin in.

Thickened antipersonnel weapons serve a different purpose. Thickening a nerve agent lessens its chances of vaporization, but increases its persistence in the environment. When an agent is thickened, its viscosity increases, allowing the chemical to contaminate areas, resist evaporation, and cause contact exposure days or weeks later. Because of the persistent nature of these agents, possible cross-contamination from one victim to another is increased, and surface penetration into equipment and clothing causes future off-gassing of the chemical.

DISPERSAL SYSTEMS

Dispersal systems used by both military and terrorist organizations have ranged from complex to extremely simple. Because of the ingenuity of the criminal mind, it is important for emergency responders to recognize that dispersal can happen anywhere, at any time, and to anyone. Chemical agents can be dispersed by aerosol delivery systems, liquid delivery systems, or sprays. Biological agents can be dispersed by introducing cultures into water systems, by aerosol dispersion, and by solid substance contamination. Nuclear agents can be dispersed by placement of a radioactive source within a structure, transportation vehicle, or public assembly. Dirty bombs pose a considerable threat for dispersing any of these agents. By placing chemical, biological, or nuclear agents inside or around an explosive, a terrorist can draw emergency responders into a post-blast area and contaminate them with an agent spread during a secondary blast. All bombings should be treated as if they contained other material. Monitoring by detection device prior to allowing unprotected workers to function in the area is a must.

PHYSICAL DECONTAMINATION

Physical decontamination includes removing clothing and providing a type of dry decontamination. For the most complete physical decontamination, a victim must remove all of his/her clothing. In many cases, this may not be realistic. When dealing with multiple ambulatory contaminated victims (mass decontamination), the most an emergency responder can expect to accomplish is the removal of outer clothing, leaving underwear in place. Even if the underwear is left in place, however, it is estimated that only 60%–80% of the contaminants can be removed. This figure is dependent on the persistency of the material and the removal procedure that is employed; both can directly affect the amount of material left on the skin.

The second method of removing contaminants physically is a process of dry decontamination. During dry decontamination, the agent is brushed off of the victim or removed using an adsorbing chemical like talcum powder, flour, Fuller's earth, baking soda, or dry soap powder. Protective lotions for aid in this step are becoming available. However, their degree of effectiveness is still unknown.

CHEMICAL DECONTAMINATION

Chemical decontamination includes using wet agents to remove or deactivate harmful chemical or biological contaminants. Wet decontamination, technical decontamination, and neutralization are types of chemical decontamination. They can be used alone or in conjunction with each other for the most efficient decontamination.

Wet Decontamination

Wet decontamination includes the use of soap-and-water solutions or mild alkaline solutions that, once applied, are accompanied with a washing or scrubbing action. Wet decontamination may include a several-step process in which the victim uses a soap-and-water wash followed by a rinse and progresses to another step, including a second soap-and-water wash and another rinse. The standard process of decontamination should include the removal of clothing then an initial removal of the agent by brushing or adsorption. This physical decontamination should be followed by a wet decontamination in which the victim is brought through wash/rinse stages using the appropriate solution (which may at times be a technical decontamination solution) with the goal of emerging from the decontamination process clean. For this reason, wet decontamination is sometimes referred to as a secondary process, because it logically follows a physical decontamination process. This combined method has proven to be the most effective technique in reducing the toxicity of chemical and biological agents.

Technical Decontamination and Neutralization

Technical decontamination includes the use of alkaline solutions and/or solvents that, when applied, neutralize or inactivate chemical or biological contaminants. Great care must be exercised when using these types of decontamination on victims. Neutralization decontamination with potent alkaline solutions or strong chlorinated powders should never be used on victims, because the damage suffered by these patients could be worse than if decontamination was not done at all. These strong chemical solutions are intended for use on equipment. Mild to moderate chemical solutions are used for the decontamination of individuals, as outlined in the following text box.

■ TECHNICAL DECONTAMINATION SOLUTIONS ■

Sodium hypochlorite solutions used for patient decontamination are 0.5% solutions made by placing 1 part unscented bleach in 10 parts water. Sodium hypochlorite solutions used for equipment and between patient contact are 5% solutions of standard off-the-shelf bleach.

Calcium hypochlorite solutions, 0.5%, can be used for patient decontamination and are prepared by placing 6 oz. of calcium hypochlorite into 5 gal. of water. Calcium hypochlorite, 5%, can be made by preparing a solution of 48 oz. calcium hypochlorite in 5 gal. of water.

Chloroamine solution is used for decontamination of mustard gas and V series nerve agents. This can be prepared by mixing 3 parts household bleach to 1 part household ammonia to 9 parts water. This will roughly give a 10% chloroamine solution. Vapors from this solution may be flammable and toxic.

ALL 5% SOLUTIONS ARE USED FOR EQUIPMENT AND HANDS THAT ARE PROTECTED WITH GLOVES BETWEEN PATIENT CONTACTS. 0.5% SOLUTION IS USED FOR PATIENT DECON. TECHNICAL SOLUTIONS SHOULD NOT BE USED AROUND THE EYES, OR IN CHEST OR ABDOMINAL WOUNDS.

Vesicants such as mustard gas and other persistent (thickened) antipersonnel agents have low solubility in water. The military suggests chloroamine and alcohol solution be used when patients are contaminated with mustard agents, especially thickened mustard agents. When such thickened agents are used, some type of technical solution will probably need to be used. Remember that most technical solutions, depending on strength, will cause harm to the patient being decontaminated. This tradeoff is sometimes necessary to lessen the overall injury or even save life.

Technical decontamination includes several washes with solutions that destroy or neutralize the contaminant. This is followed by an additional wet decontamination to rid the patient of harmful technical decontamination solutions. The process can become very involved, depending on the chemical contaminating the victim. Figure 8.1 outlines steps taken following contamination.

Physical and wet decontamination eliminate the majority of agents. However, because of the toxicity ranges of many antipersonnel weapons, the use of specialized prepared solutions are necessary to provide the most efficient decontamination. As previously discussed, specialized technical decontamination solutions may damage skin. They should not be used on or around the eyes or for the irrigation of abdominal or chest wounds. Great care should be given to patients presenting with traumatic injury when decontamination is necessary. Open wounds should only be irrigated using soap-and-water as the solution of choice. Surface skin may be decontaminated with technical solutions only if wounds can be protected. (This is especially important around abdominal and chest injuries, and mucous membranes such as the eyes.) Wounds that do not involve a cavity can have a hypochlorite solution introduced for the removal of the agent. The hypochlorite should be removed within five minutes by irrigation with water or suctioning.

Ideally, it would be convenient to have one universal decontamination solution that could complete the entire process. For the most part, soap-and-water solutions associated with physical decontamination complete the task. Unfortunately, there are thickened chemicals and biological agents that soap-and-water alone will not sufficiently remove. For example, petroleum products such as motor oil can be used as thickening agents. In these cases, a degreaser may have to be used during the decontamination process.

Many of the chemical-biological toxins are destroyed through hydrolysis. Military research indicates sodium hypochlorite achieves the most efficient alkaline hydrolysis. A 0.5% solution of sodium hypochlorite (household bleach and water mixture) works quite well in destroying nerve agents and is extremely effective against biological agents. Mustard agents and thickened nerve agents may require the use of powders, chloramine solutions (a mixture of household ammonia and household bleach), and hypochlorite solutions in order to effect 100% decontamination. See Table 8.1 for a listing of possible decontamination solutions.

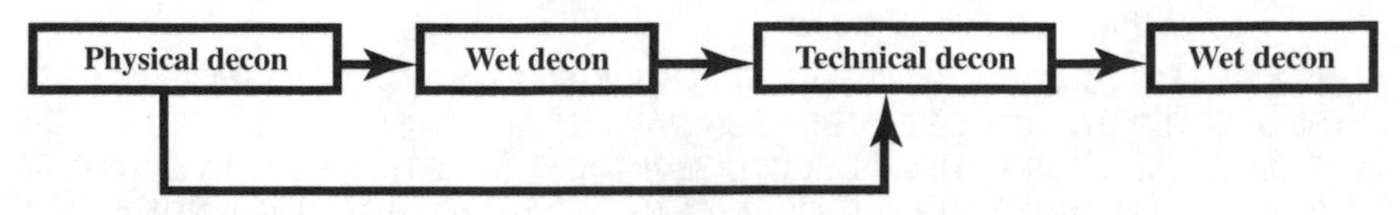

Figure 8.1 Steps Taken Following Contamination

Table 8.1 Solutions Employed in a Decontamination Corridor

Soap (perborate soap) and water	Used for wet decontamination, secondary decontamination, and after technical decontamination has been performed; Universal Decon
Sodium hypochlorite	Used as a chemical technical decontamination solution for both topical skin application (0.5%) and equipment decon (5%); works well against biologicals that do not have spore-producing capabilities and most chemical nerve agents
Calcium hypochlorite	Used to decontaminate gloved hands between patient and equipment contacts
Baking soda and baking soda solutions	Used to decontaminate patients and equipment contaminated with G series agents
Borax powder	Used to dry decontaminate chemical agents that have been thickened and biologicals
Chloramine solution	Used as a decontaminant for mustard gas and V series nerve agents
HTH, lime	Used in the shuffle pit after dry decontamination and after wet; used also to treat affected environmental surfaces

Hand and Equipment Decontamination

Between patient contacts, every effort must be made to limit secondary (cross-contamination) contamination. Rinsing gloved hands and equipment with a 5% hypochlorite solution will eliminate this problem. When dealing with mustard and V agents, a chloramine solution of 5% for hands and equipment and 0.5% solution for skin are recommended. This should immediately be followed by a soap-and-water wash.

■ MASS DECONTAMINATION CONSIDERATIONS

Action taken within the first minutes after arrival at an emergency scene will determine overall casualty outcome. Decontamination can determine the degree of medical care victims will need. Equipment common in the fire service, such as 1³/₄-inch hand lines with fog nozzles or booster lines with fog capability can work as emergency decontamination lines (Figure 8.2). These can be employed until a systematic and detailed set-up can be constructed.

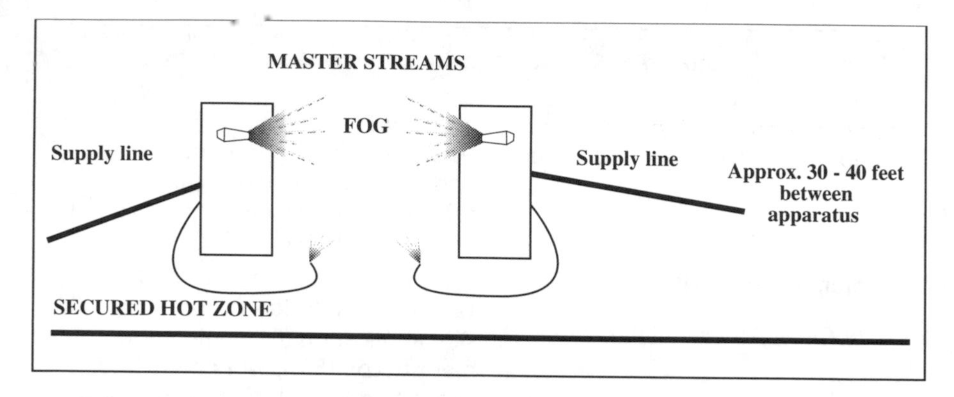

Figure 8.2 Decontamination Corridor Set-Up

Using a fire attack line, the first engine crew can begin emergency decontamination almost immediately on arrival. Full protective bunker gear with SCBA or a Level B ensemble can provide protection while decontamination is performed using a $1^3/_4$-inch or booster hose line. This procedure should be used on ambulatory patients as they approach the fire apparatus. (Nonambulatory patients would be retrieved later, as resources and protective equipment became available.)

Once resources arrive on the scene, additional fire apparatus can be positioned to create a decontaminating water curtain. By positioning two fire engines parallel to one another and approximately 30–40 feet apart, with a master stream fog device off of each engine directed toward the center, a large water curtain can be quickly established. (A supply-line water source must be secured prior to use of the master streams.) Initially it will take the first arriving units to establish this ambulatory "wash" area utilizing hand hose lines and master stream fog nozzles. As more apparatus and human resources become available, the corridor shown in Figure 8.2 can be established.

Runoff could become a problem to the workers and the environment if not addressed early on. Every effort should be made to control runoff so that a "safe" area does not become a "hazardous area," interfering with the decontamination process, as large volumes of water will be initially used. This can be done by establishing the decontamination corridor in an uphill, upwind position. Damming and isolating the area and controlling sewer and storm water drainage is a priority. The resulting contamination control area (CCA) is within the established hot zone toxic free area (TFA), but away from the vapor hazard area (VHA).

The amount of pressure that is employed to establish the water curtain should be carefully monitored so as not to produce a heavy stream of water, but rather, a large shower area for the decontamination of large numbers of contaminated patients. As resources arrive on scene, a secondary water curtain can be employed to further this decontamination process. This can enable the responders to decontaminate more patients or start the technical decontamination processes if need be.

Technical solutions can be placed in containers and inducted through existing foam lines in order to mass decontaminate. After technical decontamination washes are provided, a secondary soap-and-water wash must be established in order to remove the technical solution from the victims' bodies.

Transition to the Establishment of a Defined Decon Line

As more resources become available, a structured decontamination corridor should be established and controlled. Once the structured decon line is established, the initial emergency wash area can be abandoned or augmented to provide higher-level decontamination. At the outermost initial wash area, a secured hot zone isolation perimeter is established. Two access control points (ACP) are identified, one for the movement of patients and the other for the movement of first responders. The patient access point will identify ambulatory and nonambulatory patients who will be assigned to decontamination corridors.

Nonambulatory patients will move down one decon line, while ambulatory patients and rescuers initially on scene will move down another. The nonambulatory line will need more personnel than the ambulatory line. At the beginning of each line, a clothes disposal area will be identified and all items discarded and tagged. (This must be coordinated with law enforcement, as clothes may be considered evidence.) The victims will move in logical sequence from a disrobing area, through dry decontamination, into wet decontamination. The dry decontamination area, as well as an area prior to decontamination evaluation, should have a corridor of chlorinated lime powder or solution—a shuffle pit—where contamination carried on the feet or shoes will be destroyed prior to leaving the decontamination line.

Emergency medical care can start only after complete decontamination has occurred. This is sometimes referred to as the first level of medical treatment or 1E. At this point, detection equipment should be used to determine if the patient is clean. Once cleared from the decontamination detection area, a medical support field station can further evaluate and render emergency medical care. Specially trained Disaster Medical Assistance Team (DMAT) units would be ideal for this. If DMAT response is limited, a supportive medical team must be established. A field hospital will ultimately alleviate the need for transportation of the decontaminated patient. This reduces the possibility of moving a contaminated patient who was not evaluated properly to the medical facility. Transportation of chemically decontaminated patients will increase the supportive needs of the operation and may create other issues that are not easily remedied. Transportation to the hospital, as a rule, should be accomplished by ground vehicle. The use of helicopters is not encouraged and should be carefully analyzed. As can be seen, the decontamination corridor at the scene will require a large support staff to manage the logistics of equipment and human resources.

Hospitals face the probability of contaminated patients either walking in or being carried in by other individuals. In these cases, damage and contamination may occur before any prior knowledge of an incident exists. This will be especially true if the agent used is biological. All receiving hospitals must be able to recognize the effects of all warfare agents and hazardous materials commonly used in the community. To be fully prepared, hospitals should have available PPE, decontamination equipment (including technical decontamination solutions), isolation areas, and a trained support staff to ensure isolation from the rest of the hospital. Fire departments and field units should not be a part of the hospital's emergency plan, as these responders will be at the scene and unavailable for support roles at the medical facility. Until the contaminant is identified and controlled, hospitals must practice high-level body substance isolation and zoning control, and must notify other health services within the community. This is extremely important when dealing with biological agents. A significant event would necessitate the involvement of multiple hospitals, local health departments, community government, state agencies, and federal support, and would severely stress the medical infrastructure.

At every point where decontamination is occurring, runoff will present a problem. All runoff must be captured and regarded as toxic until evaluation and neutralization can occur. All equipment and the environment will require decontamination. The military should provide guidance for the appropriate application and disposal of decontamination solutions.

SUMMARY

Anyone who has set up a decontamination corridor knows that this is a long and laborious job, requiring vast human resources, set-up logistics, and equipment. It is for these reasons and many others that HazMat teams, along with the new MMSTs, have been developing decontamination trailers for immediate set-up at a hazardous materials incident or terrorist strike. If you are considering these trailers for decontamination or mass decon, here are some considerations for implementation:

- Construct the trailer so that minimal set-up and human resource is required. Initial set-up should occur in less than five minutes using two to three responders.
- The trailer should be designed to decontaminate the usual number of personnel at a hazardous materials incident, as well as mass decontamination (approximately 300–500 victims per hour).
- Interior sections should incorporate the following: cascade system and hard-line system for responders within the corridor; climate control in order to support total gross decontamination; modesty barrier, a male corridor with male operators, and a female corridor with female operators; limited solution application nozzles/showers with containment area; lighting inside and outside to accommodate night-time operations; an induction system by which decon solutions could be applied when required; water heating capability; and air exchangers within the trailer capable of scrubbing the air and HEPA filters for biologicals.

Decontamination is a vital part of an entire operation. It will be challenging both functionally and logistically. Insuring that victims will receive decon in a timely, functional manner will be the ultimate challenge for any response agency. Figure 8.3 is an overview of a decontamination corridor setup.

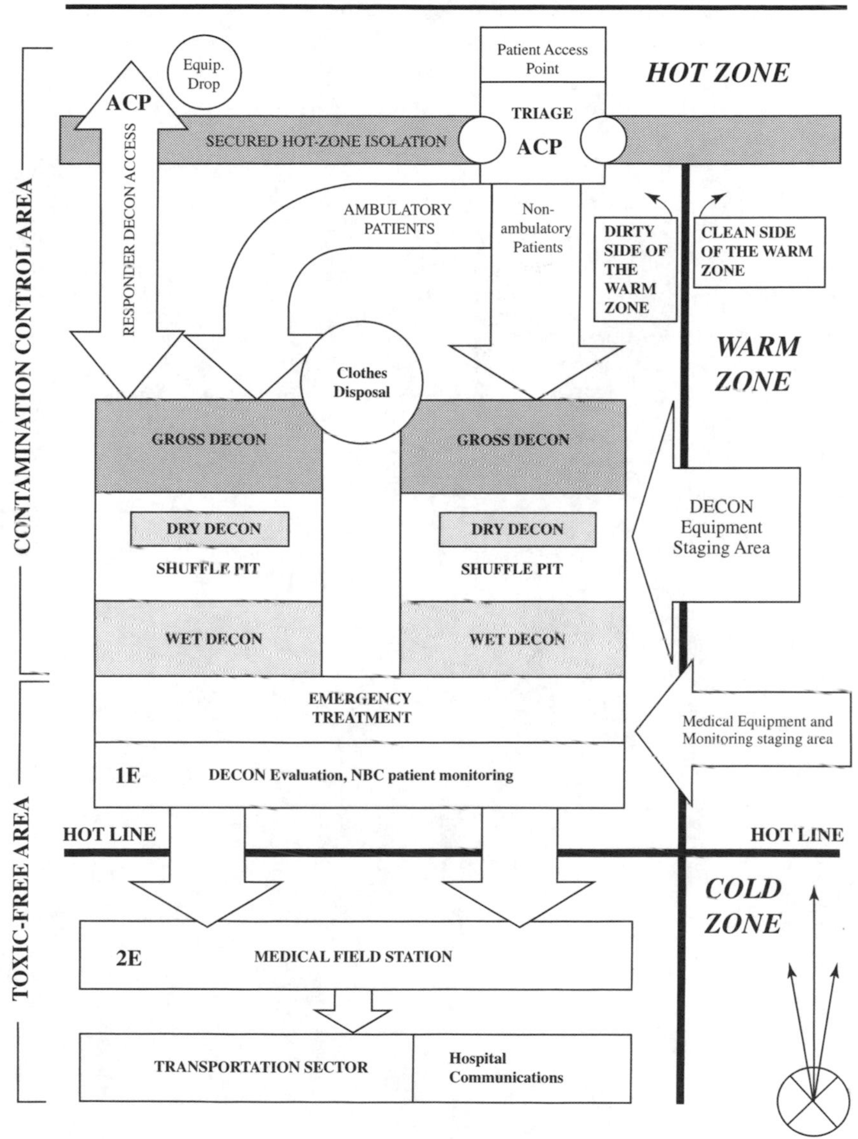

Figure 8.3 Decontamination Corridor Set-Up

APPENDIXES

- A Target and Signature Profile
- B NBC Operational Preparedness Matrix
- C-1 Neurotoxins Informational Card
- C-2 Blister Agents Informational Card
- C-3 Chemical Asphyxiants Informational Card
- C-4 Respiratory Irritants Informational Card
- D Toxicological Terminology
- E Decontamination Solutions: Use and Preparation
- F Colorimetric Tubes and Their Detection Characteristics
- G PPE
- H Complete Decon Corridor
- I START Triage Method

Appendixes A and B have two important functions: to help analyze the incident and to help identify the chemical.

The Target and Signature Profile enables the responder to identify clues to historically common targets. Included are signature profiles that have occurred or have the potential to occur given the right set of circumstances. This analysis is divided into five groups. The first three identify the event, while the last two are on-scene considerations. "Location or occupancy," "primary event type," and "timing of the event" are all interrelated. "Location or occupancy" examines historical context and targets, including representative organizations or unusual context. These are organiztions or gatherings that may be offensive to the extremist group. Large areas also provide an opportunity for high-impact, media-driven events. "Primary event type" analyzes how calls are sometimes received; a responder should be highly suspicious until events are proven one way or another. "Scene size-up" and "scene management" list considerations and resources that should ben analyzed by the first responding units and the incident commander.

The NBC Matrix organizes information presented in the text. The chemicals list will enable the reader to find important information within a matter of seconds. A more detailed description of the first four categories is given in Appendix C.

Appendix A Target and Signature Profile

LOCATION OR OCCUPANCY

- ☐ High Profile Owner/Operator
- ☐ Controversial Business
- ☐ Public Buildings
- ☐ Infrastructure
  - ☐ Utilities
  - ☐ Public Services
  - ☐ Transportation Routes
- ☐ Assembly Areas
  - ☐ Public
  - ☐ Private
  - ☐ Public Transportation
- ☐ Symbolic Target
  - ☐ Religious
  - ☐ Event Venue
  - ☐ Government
- ☐ High-Profile Media Event
- ☐ Political Environment
  - ☐ Political Charged
  - ☐ Diplomat
- ☐ Suspicious Package
- ☐ Multiple Request for Help
- ☐ Historical Target
- ☐ Educational Establishment

PRIMARY EVENT TYPE

- ☐ Explosion
  - ☐ Explosion Without High Degree of Damage
  - ☐ Explosion with Mist or Vapor Production
  - ☐ Explosion with Unusual Liquid or Fog
  - ☐ Unusual Increases or Decreases in Number of Animals and Insects
  - ☐ Level of Explosion Severity
    - ☐ Minimal ☐ Moderate ☐ Severe
- ☐ Epidemiology Defines Patterns of Illness
  - ☐ Unusual Multiple Casualties
    - ☐ Trauma
    - ☐ DUMBELS
    - ☐ Patterns of Illness
  - ☐ Unexplained Plume, Gas, Vapor

TIMING OF THE EVENT

- ☐ Historical Context
- ☐ Cultural and Social Beliefs
- ☐ Holiday
- ☐ Demographics
- ☐ Time of Day/Year

SCENE SIZE-UP

- ☐ Items That Seem Out of Place
- ☐ Unexplained Signs and Symptoms
- ☐ DUMBELS Effect
- ☐ Response to Assistance with Defined Patterns
- ☐ Weather and Wind Speed/Direction
- ☐ Plume Direction If Fog, Vapors, Gases Present
- ☐ Unexplained Odors, Tastes from Odor
- ☐ Higher Than Normal Number of Respiratory Illnesses
- ☐ Structural Damage

SCENE MANAGEMENT

- ☐ Prepare for Secondary Device, Isolate Radio Transmissions
- ☐ Plan for Protective Equipment During Emergency Scene Activity
- ☐ Isolate Entire Area—Confine Incident
- ☐ Evacuate and Isolate Victims Within Inner Hot Zone
- ☐ Protect Victims Within the Outer Cold Zone
- ☐ Prepare for Mass Decontamination
- ☐ **NOTIFICATIONS**
  - ☐ Local Emergency Resources
  - ☐ Hospitals
  - ☐ Health Department
  - ☐ ATF
  - ☐ National Guard
  - ☐ State/County Warning Point
  - ☐ Regional HazMat Teams
  - ☐ U.S. Public Health
  - ☐ Bomb Squad
  - ☐ EOC
  - ☐ Mutual Aid
  - ☐ CDC
  - ☐ MMST
  - ☐ FBI, MBI
  - ☐ Police, DEA
  - ☐ FEMA
  - ☐ DOD—Army Chem/Bio
  - ☐ DOE

Appendix B NBC Operational Preparedness Matrix

TYPICAL MILITARY NBC AGENTS		EXPOSURE LIMITS		DETECTION METHOD			
		TWA ppm	ICt_{50}/LCt_{50} mg-min/m3	Colorimetric	M-8	M-9	M256
NEUROTOXINS (Nerve Agents) specific organophosphorus compounds that bind with acetylcholinesterase							
TABUN	GA	.00001	300/400	Phosphoric acid esters	YELLOW Nonpersistant G agents OD/BLACK Persistant V agent	Red, Brown, Purple, Pink Dependant on conc. + interference	Tube Two reaction (T-401)
SARIN	GB	.00002	75/100				
SOMAN	GD	.000004	Unk./70				
TABUN (derivative)		.00001	Unk./Unk.				
V agents	VX	.0000009	35/50	UNK.			
BLISTER AGENTS (Vesicants) cause severe chemical burns to the skin, mucus membranes, and respiratory system							
Sulfur MUSTARD	HD	.0003-.0004	200/1500	Thioether	RED	Red, Brown, Purple, Pink—Dependant	T-402
Nitrogen MUSTARD	HN	.0003-.0004	200/1500	Organic nitrogen			UNK
LEWISITE	L	.0003	300/1000-1500	Organic arsenic/arsine			
PHOSGENE OXIME	CX	.086	UNK.	Phosgene	UNK.		T-403
RESPIRATORY IRRITANTS (Choking Agents) pulmonary irritation that produces copious pulmonary fluid and thus asphyxiation							
CHLORINE	CL	0.5	15-30 ppm/40-60 ppm	Chlorine	No military detection. Colorimetric tubes and standard detection		
PHOSGENE	CG	0.1	3-5 ppm/50-80 ppm	Phosgene			
CHEMICAL ASPHYXIANTS (Blood Agents) chemical asphyxiation due to chemical bonds within the cell interfering with cellular aerobic metabolism							
HYDROGEN CYANIDE	AC	4.7	Dependant on conc./2500-5000	Hydrogen cyanide/ hydrocyanic acid	No military detection. Colorimetric tubes and standard detection		
CYANOGEN CHLORIDE	CK	0.3	2500/11000	Cyanogen chloride			
LACRIMATORS and VOMIT AGENTS		Dependant on the agent		Dependant on the agent	No military detection. Standard detection		
BACTERIA, BIOTOXINS, and VIRUSES.		Latency dependant on the concentration and organism		NONE to date. Laser analysis and immunoassays are currently being developed.			
NUCLEAR ISOTOPES		Dependant on the isotope, dispersion method, and distance from the source		Standard radiation detection inclusive of scintillation and luminescent devices. Military notification and NEST.			

TYPICAL MILITARY NBC AGENTS	SIGNS AND SYMPTOMS	DECON SOLUTIONS	TREATMENT	PPE
NEUROTOXINS (Nerve Agents) specific organophosphorus compounds that bind with acetylcholinesterase				
TABUN GA SARIN GB SOMAN GD TABUN (derivative)	D-Diarrhea U-Urination M-Miosis B-Bronchospasms E-Emesis L-Lacrimation S-Salivation See page XX for details	Sodium (calcium) hypochlorite, 0.5% sol. followed by a soap-and-water wash, followed by a soap/water wash	Decon using the appropriate solutions. Oxygenate ASAP. IV 2–5mg atropine q 5 min IV 1 gm pralidoxime over 2 min Valium IV for seizures Monitor patient See page 25 for details	Nonpersistant LEVEL A (OSHA APPROVED) Persistant LEVEL C with M17 mask (NOT OSHA APPROVED)
V agents VX		Follow blister agent decon below		
BLISTER AGENTS (Vesicants) cause severe chemical burns to the skin, mucus membranes, and respiratory system				
Sulfur MUSTARD HD Nitrogen MUSTARD HN LEWISITE L	See page 34 for detailed signs and symptoms	Dichloramine solution followed by sodium (calcium) hypochlorite, .5% and soap-and-water wash HD decon before pain	There is no field treatment for blister agents beyond good decontamination. Decontaminate using appropriate solutions.	LEVEL A and B (OSHA APPROVED) LEVEL C with M17 mask (NOT OSHA APPROVED)
PHOSGENE OXIME CX		Copious soap-and-water after pain		
RESPIRATORY IRRITANTS (Choking Agents) pulmonary irritation that produces copious pulmonary fluid and thus asphyxiation				
CHLORINE CL PHOSGENE CG	See page 32 for detailed signs and symptoms	Copious soap-and-water wash	Decon using appropriate solutions. See page 33 for detailed treatment	Level A, B, C or C with APR dependant on concentrations
CHEMICAL ASPHYXIANTS (Blood Agents) chemical asphyxiation due to chemical bonds within the cell interfering with cellular aerobic metabolism				
HYDROGEN CYANIDE AC CYANOGEN CHLORIDE CK	See page 28 for detailed signs and symptoms	Copious soap-and-water wash	Decon using appropriate solutions. See page 29 for detailed treatment.	Level A, B, C or C with APR dependant on concentrations
LACRIMATORS and VOMIT AGENTS	See page 37 for detailed signs and symptoms	Copious soap-and-water wash	Decon using appropriate solutions. See page 37 for detailed treatment.	Level D
BACTERIA, BIOTOXINS and VIRUSES	Dependant on agents used	See Neurotoxin decon	Dependant on agent.	Level A, B, C or C with APR dependant on type
NUCLEAR ISOTOPES	Dependant on isotope strength, time, distance, and shielding	Copious soap-and-water wash	Decon using appropriate solutions. Supportive treatment.	Time, distance, shielding

Appendix C-1 Neurotoxins Informational Card

NEUROTOXINS — (Nerve Agents) Specific organophosphorus compounds that bind with acetylcholinesterase

SIGNS AND SYMPTOMS

Nose—Rhinorrhea
Eyes—Constricted pupils, lacrimation, conjunctivitis (dim, blurred vision, pain)
Respirtory—Hypersecretion production, bronchospasms, (dyspnea, cough)
Gastrointestinal—Cramps, diarrhea, vomiting, hypermotility
Somatic—Weakness, fasciculations, seizures
CNS—Anxiety, restlessness, coma, respiratory and circulatory depression
Other—Salivation, sweating
DUMBELS

BASIC TREATMENT

1. Remove victim from environment
2. Decontaminate immediately
3. Open and maintain airway
4. High-flow oxygen ASAP
5. Continue with decontamination
6. Consider advanced treatment if available and/or prepare for transport
7. Monitor and suction patient
8. Consider secondary contamination and appropriate PPE

ADVANCED TREATMENT

1. Procedures as identified in the basic treatment protocol
2. Evaluate oxygenation, high-liter flow before ALS
3. 2–5 mg atropine IVP q 5 min
4. 1 gm pralidoxime over 2 min
5. Valium for seizures
6. Monitor patient during transport

EXPOSURE LIMITS

TABUN	GA	.00001 ppm
SARIN	GB	.00002 ppm
SOMAN	GD	.000004 ppm
TABUN (derivative)	GF	.00001 ppm
V Agents	V	.0000009 ppm

DETECTION METHOD

Colorimetric tubes—Phosphoric acid esters for G agents
M-8 detection paper—Yellow indicates a nonpersistent G agent OD/black indicates persistent V agent
M-9 detection paper—Red, brown, pink, dependent on the concentration and interference solutions or gases
M256 Tube Two reaction indicates nerve agent present
FID/PID eV is less then 10.6; unknown relative reaction ratios
ICAM is useful

PHYSICAL PROPERTIES

Clear, colorless, and odorless
Liquid form with low vapor pressure
Thicking agents are sometimes applied to increase persistency
Liquid may easily be aerosolized

DECONTAMINATION

1. Remove from contaminated area
2. Gross decontamination
3. Sodium (calcium) hypochlorite, 0.5% solution
4. Degreaser if petroleum product used for persistency
5. Follow by a soap-and-water wash
6. Evaluate decontamination

PPE

Level A within a neurotoxin environment
Level B during decon
Level C if persistent agent with M17 (military APR) mask or appropriate APR (NOT OSHA APPROVED)

Appendix C-2 Blister Agents Informational Card

BLISTER AGENTS — (Vessicants) cause severe chemical burns to the skin, mucus membranes, and repiratory system

SIGNS AND SYMPTOMS

SKIN—Blisters, pain, burning
PULMONARY—Irritation, airway obstruction, pulmonary edema
OCULAR—Pain, blepharospasm, irritation
See Chapter 3 for details between the agents

BASIC TREATMENT

1. Remove victim from environment
2. Decontaminate immediately
3. Maintain an open airway
4. Administer high-flow oxygen
5. Consider secondary contamination and appropriate PPE
6. Prepare for transport

ADVANCED TREATMENT

1. Procedures as identified in the basic treatment protocol
2. Thorough decontamination associated with high-level patient isolation techniques, in order to control secondary infection and cross-contamination by body fluids
3. Dimercaprol (BAL) 2–4 mg/kg q 4–12 hours

EXPOSURE LIMITS

Sulfur MUSTARD	HD	.0004 ppm
Nitrogen MUSTARD	HN	.0003 ppm
LEWISITE	L	.0003 ppm
PHOSGENE OXIME	CX	.086 ppm

DETECTION METHOD

Colorimetric tubes—Thioether for sulfur mustard; organic nitrogen for nitrogen mustard; organic arsenic/arsine for Lewisite; phosgene for phosgene oxime
M-8 detection paper—if it turns red, a blister agent, unknown if CX present
M-9 detection paper—red, brown, purple, pink, dependent on concentration, unknown if CX present
M256 will only detect sulfur mustard with a positive reaction
ICAM is useful

PHYSICAL PROPERTIES

Colorless to brown-/violet-/yellow-/amber-colored liquid
Liquids that vaporize slowly
Liquids may be easily aerosolized
Phosgene oxime is a solid that vaporizes

THESE AGENTS ARE STRONG IRRITANTS

DECONTAMINATION

1. Remove from contaminated area
2. Gross decontamination
3. Chloroamine, 10% solution wash
4. Sodium (calcium) hypochlorite, 0.5% solution wash
5. Copious soap-and-water wash
6. Evaluate decontamination

See Chapter 8 for details on solutions for specific agents

PPE

Level A within a blister agent environment
Level B during decon
Level C with M17 (military APR) mask or appropriate APR (NOT OSHA APPROVED)

Appendix C-3 Chemical Asphyxiants Informational Card

CHEMICAL ASPHYXIANTS — (Blood Agents) chemical asphyxiation due to chemical bonds within the cell interfering with cellular aerobic metabolism

SIGNS AND SYMPTOMS

Respiratory—(Early) Tachypnea, hyperpnea, dyspnea; (Late) Depressed respiratory rate, respiratory depression, apnea and death

Cardiovascular—(Early) Flushing, hypertension, reflex bradicardia, AV or intraventricular rhythms; (Late) Hypotension, acidosis, tachycardia, ST segment changes

BASIC TREATMENT

1. Remove victim from environment
2. Decontaminate immediately
3. Maintain an open airway
4. Administer high-flow oxygen
5. Prepare for transport
6. Consider secondary contamination

ADVANCED TREATMENT

1. Procedures as identified in the basic treatment protocol

FOLLOWING USED FOR CYANIDES AND H_2S ONLY

2. Amyl nitrite perles inhaled for 15–30 seconds while IV is established
3. Sodium nitrite, 300 mg per 10 cc IV slow
4. Consider positioning when blood pressure drops
5. Sodium thiosulfate, 50 ml of a 25% solution over 10 minutes (not to be used for H_2S)
6. Monitor patient during transport

EXPOSURE LIMITS

HYDROGEN CYANIDE	AC	4.7 ppm
CYANOGEN CHLORIDE	CK	0.3 ppm
HYDROGEN SULFIDE	H_2S	50 ppm
ARSINE		.05 ppm

DETECTION METHOD

Colorimetric tubes—Hydrogen cyanide/hydrocyanic acid for hydrogen cyanide; cyanogen chloride for cyanogen chloride; hydrogen sulfide for hydrogen sulfide; arsine for arsine

Electrochemical/Microprocessing detection

PID

No military detection—standard detection modalities

PHYSICAL PROPERTIES

Colorless with irritating odor

Liquids that vaporize

Liquids may be easily aerosolized

or

Gaseous state

Smell can rapidly fatigue the senses

DECONTAMINATION

1. Remove from contaminated area
2. Gross decontamination
3. Copious soap-and-water wash
4. Evaluate decontamination

PPE

Level A within a chemical asphyxiant environment

Level B during decon

Appendix C-4 Repiratory Irritants Informational Card

RESPIRATORY IRRITANTS — (Choking Agents) pulmonary irritation that produces copious pulmonary fluid and asphyxiation

SIGNS AND SYMPTOMS

1. High solubility with water—injury will involve the upper respiratory system, bronchospasms, laryngeal spasms; localized chemical burns, irritation, upper airway swelling, and occlusion
2. Low solubility with water—injury will be deeper in the airways affecting the fine bronchioles and alveoli; auscultation will reveal rales with the probability of noncardiogenic pulmonary edema

BASIC TREATMENT

1. Remove victim from environment
2. Decontaminate immediately
3. Maintain an open airway
4. Administer high-flow oxygen
5. Consider secondary contamination and appropriate PPE
6. Prepare for transport

ADVANCED TREATMENT

1. Procedures as identified in the basic treatment protocol
2. Upper airways dilated using updrafts of alupent or albuterol
3. Further dilation by using brethine or epinephrine SQ
4. PEEP ventilation for lower airway injury with PE
5. Monitor patient during transport

EXPOSURE LIMITS

CHLORINE	CL	0.5 ppm
PHOSGENE	CG	0.1 ppm
AMMONIA	NH_4	25 ppm

DETECTION METHOD

Colorimetric tubes—Chlorine for chlorine; phosgene for phosgene; ammonia for ammonia
Electrochemical/Microprocessing Detection
PID
No military detection—standard detection modalities

PHYSICAL PROPERTIES

Liquids that vaporize extremely easily

THESE AGENTS HAVE STRONG IRRITATING QUALITIES

DECONTAMINATION

1. Remove from contaminated area
2. Gross decontamination
3. Copious soap-and-water wash
4. Evaluate decontamination

PPE

Level A within a respiratory irritant environment
Level B during decon
Level C with M17 (military APR) mask or appropriate APR (NOT OSHA APPROVED)

Appendix D Toxicological Terminology*

TLV-TWA	An 8-hour day, 40-hour work week with repeated exposure without any adverse effects, followed by a 60-minute break, not to exceed 4 times a day
TLV-STEL	Fifteen-minute excursion in which the worker is exposed to the chemical continuously. Must not have any of the following effects: 1. irritation 2. chronic tissue damage 3. the impairment of a self-rescue
TLV-EL	An average exposure not to exceed 5 times the published 8-hour limit
TWA	This will not occur for more than 30 minutes on any work day
TLV-s	Identifies a material that is absorbed through the skin
TLV-c	Identifies a ceiling level
IDLH	The maximum airborne contamination that an individual could escape from in 30 minutes without any side effects
PEL	Same as TLV-TWA
REL	Same as TLV-TWA
LC_{Lo}	The lowest concentration of airborne contaminants that cause injury
LD_{Lo}	The lowest dose (solid/liquid) that cause an injury
LC_{50}	50% of the test population died from the introduction of this airborne contaminant
LD_{50}	50% of the tested population died from the introduction of this chemical, which may be a solid, liquid, or gas
LCt_{50}	A statistically derived LC_{50} (LDT_{50} is a statistically derived lethal dose) LCt_{50} lethal concentration of 50% of the population; based on time and dose
MAC	Maximum allowable concentration
RD_{50}	Respiratory depression of 50% of the observed population, 50% calculated concentration of respiratory depression, toward an irritant, over a 10–15 minute time frame
ICt_{50}	Incapacitating concentration of 50% of the population
mg-min/m^3	Concentration over time within a defined volume of a meter cubed. This is a relative constant, even with a difference in concentration and time (Harper's Law). Example: 4 mg-min/m^3 for 10 minutes is 40 mg/m^3, so is 8 mg-min/m^3 for 5 minutes. See LCt_{50}.

* Abbreviations commonly found within current literature. There are other abbreviations within the discipline of toxicology. All are based on testing procedures and current scientific study. All numerical expressions should be well understood before applying them to the hazardous materials or terrorist event.

Appendix E Decontamination Solutions: Use and Preparation

Decontamination Solution	Uses	Preparation	Remarks
Patients			
Soap and Water	Flush rinse, frequency-dependent	1 lb per 5 gal	Must follow all technical decontamination procedures; may require several rinses.
Sodium hypochlorite	Chem-bio G, V agents blister agents	1 part bleach to 10 parts water makes 0.5% 2 parts per 10 for bio	Skin and respiratory hazard, corrosive qualities. 15-min or greater contact time is suggested for nonspore biologicals.
Calcium hypochlorite	Chem-bio G, V agents blister agents	5 lbs in 12 gal water for 5% (equip) 0.5 lbs in 12 gal of water for 0.5% (person)	Can react with a variety of chemicals. Skin and respiratory hazard; extensive flushing required. Will give off toxic vapors once in contact with G agents. VX and HTH will burn. 15-min contact for spore-forming biologicals.
Chloramine	Mustard V agents	3 parts bleach to 1 part household ammonia in 9 parts water for roughly a 10% solution	Works well against mustard gas exposure. Skin and repiratory hazard; extensive secondary decontamination using soap-and-water wash is necessary.
Equipment			
Ammonia	Chem-bio G agents chem	Concentration as available in household	Skin and repiratory hazard; extensive flushing required. Effective on nonspore bios.
Acetone (2-propanone)	G, V agents Blister agents	Concentration as available in household	FLAMMABLE. Does not neutralize the agent, only flushes; runoff is a concern. May cause hypothermia and toxicity if used on patients in cold environment.
Diethyl ether	Chem G, V agents Blister agents	Concentration as available	Same as acetone (2-propanone)
Dichloroamine	Mustard	Decon powder in organic solvents to achieve a 10% solution	Skin and repiratory hazard; extensive flushing required. May be effective on nonspore bios. May be flammable and toxic. Skin, repiratory, and eye exposure should be avoided.
Ethylene glycol	Chem	Concentration as available	Does not neutralize the agent; all runoff must be controlled.
Sodium hydroxide	Chem-bio Neutralizes G agents, effective on L and spore bios	10 lbs to 12 gal water for a 10% solution mixed in steel container	Skin, respiratory, and eye exposure should be treated with L-A PPE. Runoff is toxic and corrosive. Areas in contact with this solution should be flushed with vinegar (diluted acetic acid).
Potassium hydroxide	Same as above	Same as above	Same as above
Sodium carbonate	Chem G agents	10 lbs in 12 gal water for a 10% solution	Hazardous products with VX combination

Appendix F Colorimetric Tubes and Their Detection Characteristics

AGENT	TUBE	DETECTION RANGE
Tabun	Phosphoric acid esters	0.05 ppm
Sarin	Phosphoric acid esters	0.05 ppm
Soman	Phosphoric acid esters	0.05 ppm
Chloropicrin	Carbon tetrachloride	1–15 ppm
Lewisite	Organic arsenic compounds	3 mg/m^3
	Arsine	0.1 ppm
Mustard	Thioether	1 mg/m^3
Nitrogen mustard (HN-1)	Organic basic nitrogen compounds	1 mg/m^3
Nitrogen mustard (HN-2)	Organic basic nitrogen compounds	1 mg/m^3
Nitrogen mustard (HN-3)	Organic basic nitrogen compounds	1 mg/m^3
Hydrogen cyanide	Hydrogen cyanide	2–30 ppm
	Hydrocyanic acid	2–30 ppm
Chlorine	Chlorine	0.2–30 ppm
Phosgene	Phosgene	0.02–1 ppm
Arsine	Arsine	0.05–3 ppm
Cyanogen chloride	Cyanogen chloride	0.25–5 ppm
Pepper spray	Olefins (NOT RECOMMENDED)	1–55 mg/l
Mace	Chloroformates	0.2–10 ppm
	Acid aldehydes (NOT RECOMMENDED)	100–1000 ppm

(Courtesy of Drager Corporation, provided through their civil defense branch)

Appendix G PPE

The following is a list of example ensembles that give appropriate protection to the responder. Further referencing is encouraged.

AGENT LEVEL OF CONCERN	LEVEL A ENTRY	LEVEL B DECON
Tabun	Trelleborg polyethlene, tychem	Saranex with SCBA
Sarin	Trelleborg polyethlene, tychem	Saranex with SCBA
Soman	Trelleborg polyethlene, tychem	Saranex with SCBA
Chloropicrin	Trelleborg teflon, responder	Saranex with SCBA
Lewisite	Trelleborg polyethlene, tychem	Saranex with SCBA
Mustard	Trelleborg polyethlene, tychem	Saranex with SCBA
Nitrogen mustard (HN-1)	Trelleborg polyethlene, tychem	Saranex with SCBA
Nitrogen mustard (HN-2)	Trelleborg polyethlene, tychem	Saranex with SCBA
Nitrogen mustard (HN-3)	Trelleborg polyethlene, tychem	Saranex with SCBA
Hydrogen cyanide	Trelleborg teflon	Saranex with SCBA
Chlorine	Trelleborg butyl rubber, neoprene, teflon, viton	Saranex with SCBA
Phosgene	Trelleborg butyl rubber, neoprene, teflon, viton	Saranex with SCBA
Arsine	Trelleborg polyethlene, tychem	Saranex with SCBA
Cyanogen chloride	Trelleborg polyethlene, tychem	Saranex with SCBA
Pepper Spray	Saranex with SCBA	Level C
Mace	Saranex with SCBA	Level C

Appendix H Complete Decon Corridor

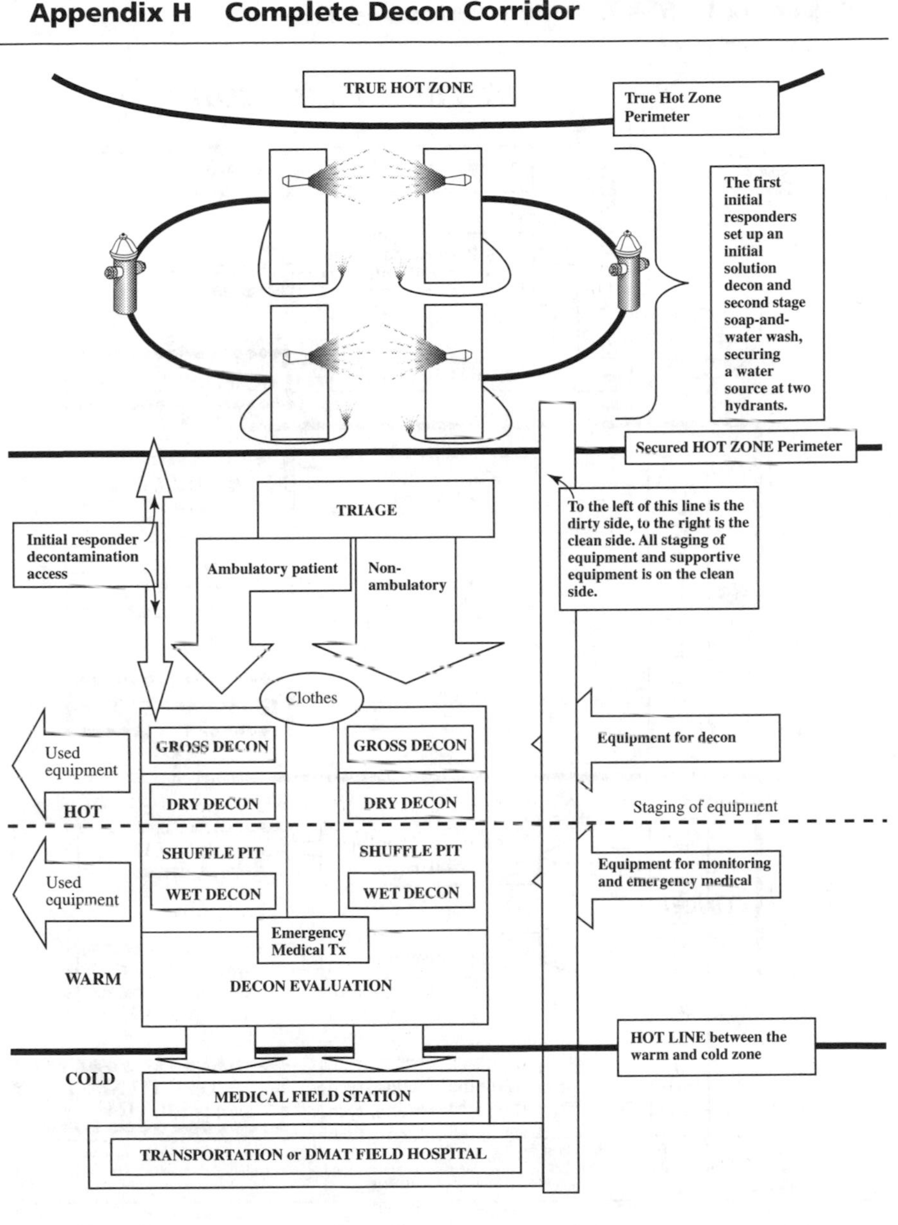

Appendix I START Triage Method

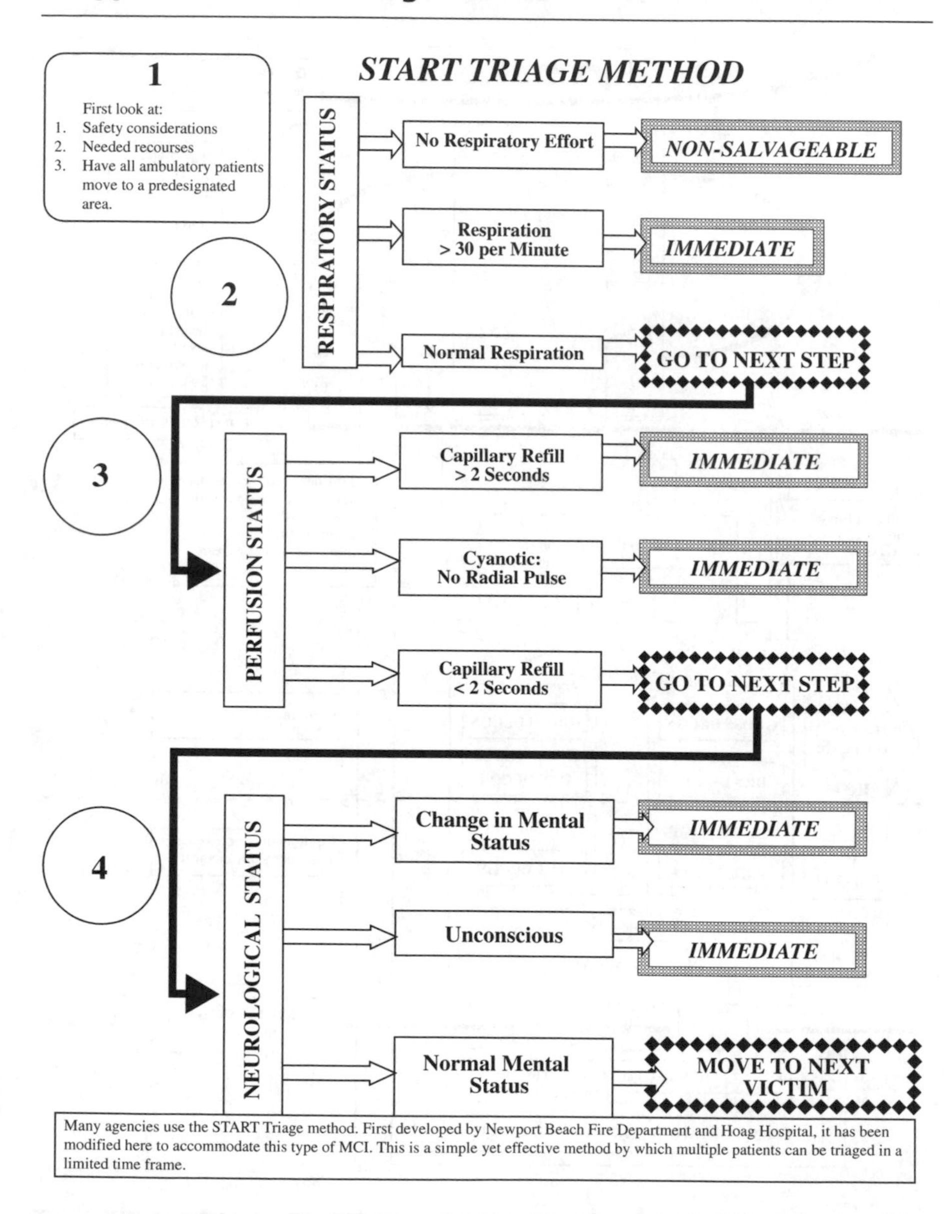

Many agencies use the START Triage method. First developed by Newport Beach Fire Department and Hoag Hospital, it has been modified here to accommodate this type of MCI. This is a simple yet effective method by which multiple patients can be triaged in a limited time frame.

GLOSSARY OF ACRONYMS AND TERMS

The jargon used within the emergency response field, fire service, law enforcement, and military consists of words formed by the initial letters of each word in a phrase and acronyms assigned to particular definitions. Not all of the following are discussed in this handbook. However, all are used within the disciplines previously mentioned. There are many more. We have chosen the most common and those that are appropriate for this discussion.

Acronyms

1E	First Level of Medical Treatment
2E	Initial Professional Medical Care (Hospital or Field Unit)
AC	Hydrogen cyanide
APR	Air Purifying Respirator
ATSDR	Agency for Toxic Substances and Disease Registry
ACP	Access Control Points
ALS	Advanced Life Support
APR	Air-purifying Respirator
ATF	Federal law enforcement agency: Alcohol, Tobacco and Firearms
CA	Bromobenzylcyanide—LACRIMATOR AGENT
CAM	Chemical Agent Monitor
CCA	Contamination Control Area
CDC	Center for Disease Control
CG	Phosgene
CGI	Combustable Gas Indicator
CHEMTREC	Chemical Transportation Emergency Response Center
CHF	Congestive Heart Failure
CK	Cyanogen chloride
CL	Chlorine
CN	2-chloroacetophenone—LACRIMATOR AGENT
CNC	Chloroacetophenone in chloroform—LACRIMATOR AGENT
CNS	Chloroacetophenone and chloropicrin in chloroform—LACRIMATOR AGENT or Central Nervous System
CO	Carbon Monoxide
CR	Dibenoxazepine—LACRIMATOR AGENT

CRC	Contamination Reduction Corridor—same as Warm Zone or Decontamination Corridor
CRZ	Contamination Reduction Zone—Decontamination Corridor
CP	Command Post
CS	O-chlorobenzylidenemalononitrile—LACRIMATOR AGENT
CX	PHOSGENE OXIME—dichloroformoxine-CCL2NOH
DA	Diphenylchloroarsine—VOMIT AGENT (Riot Control Agent)
DC	Diphenylcyanoarsinine—VOMIT AGENT (Riot Control Agent)
DM	(Adamsite)—Diphenylaminochloroarsine—VOMIT AGENT (Riot Control Agent)
DOT	Department of Transportation
DEA	Drug Enforcement Agency
DMAT	Disaster Medial Assistance Team
DOD	Department of Defense
DOE	Department of Energy
ED	Ethyldichloroarsine—blister agent
EHS	Extremely Hazardous Substance
EKG	Electrocardiogram—Electrical wave forms produced by the heart. Waves are called P, Q, R, S, and T.
EMP	Electromagnetic Pulse
EMS	Emergency Medical Systems—emergency medical providers
EPA	Environmental Protection Agency
EOC	Emergency Operations Center
EZ	Exclusion Zone—Hot Zone
FBI	Federal Bureau of Investigation
FCAM	Field Chemical Agent Monitor
FEMA	Federal Emergency Management Agency
FID	Flame Ionization Device (detector)
G	G-Agent or German nerve agent
GA	Tabun—ethyl N,N-dimethylphosphoroamidocyanidate
GB	Sarin—isopropyl methylphosphonofluoridate
GD	Soman—pinacolyl methylphosphonofluoridate
GF	Tabun form—cyclohexyl methylphosphonofluoridate
HEPA	High Efficiency Particulate Air
HD	Mustard gas—[bis-(2chloroethyl)sulfide]
HN, HN_2, HN_3	Nitrogen Mustard (HN_2 mechlorethamine)
IC	Incident Commander
ICAM	Improved Chemical Agent Monitor
ICP	Incident Command Post
IM	Intramuscular drug injected into muscle.
IV	Intravenous drug or fluid placed into vein.
L	Lewisite—dichloro-(2-chlorovinyl) arsine
MBI	Metropolitan Bureau of Investigation
MD	Methyldichloroarsine—blister agent
MCI	Mass Casualty Incident
MMST	Metropolitan Medical Strike Team

MOPP	Mission-Oriented Protective Posture
MSD	Minimum Safe Distance
NBC	Nuclear, Biological and Chemical
NDMS	United States Public Health Service
NEST	Nuclear Emergency Search Teams
NRC	National Response Center
OSHA	Occupational Safety and Health Administration
PID	Photoionization Device (detector)
RAD	Radiation Absorbed Dose
REM	Roentgen Equivalent Man (biological effects)
SCBA	Self-Contained Breathing Apparatus
SOP	Standard Operating Procedures
SWAT	Special Weapons and Tactics
SZ	Support Zone—Cold Zone
T2	Tricholthecene mycotoxin
TCON	Threat Conditions; A low; B medium, C high, D imminent
TFA	Toxic-Free Area
TNT	Trinitrotoluene
OC	Oleoresin Capsicum
OSC	Federal On-Scene Commander
PD	Variant of Lewisite—phenyldichloroarsine
PDD-39	Presidential Decision Directive
PPE	Personnel Protective Equipment
P-Tab	Pyridostigmine Bromide Tabs
TLV	Threshold Limit Value
TWA	Time Weighted Average
USAR	Urban Search and Rescue
V	Venom, used to describe a class of nerve agent
VEE	Venezuelan Equine Encephalitis
VHA	Vapor Hazard Area
VHF	Viral Hemorrhagic Fever
VX	V agent—O-ethyl S-(2-diisopropylaminoethyl) ethylphosphonothiolate

Glossary of Terms

Absorption The incorporation of a material into another, or the passage of a material into and through the tissues. The ability of a material to draw within it a substance that becomes a part of the original material.

Acetylcholinesterase An enzyme that hydrolyzes (washes) acetylcholine.

Acid Chemicals with a hydrogen ion concentration and pH of less than 7.

Acute Refers to short term. Can be used to explain either a short-term exposure or the rapid onset of symptoms.

Adsorption A process by which a substance will take up and hold a gas, liquid, or dissolved substance on its surface.

Aerosol A finely divided liquid or solid in gas or vapor suspension.

Air purifying respirator (APR) Filtration mask that filters outside air.

Alkali A chemical with a concentration of hydronium ions and a pH of greater than 7.

Alveoli Microscopic air sacs in the lungs.

Alveolar wall The semipermeable membrane that makes up the walls of the alveoli.

Asphyxiant A chemical that displaces oxygen in the air or interferes with the use of oxygen in the body. An asphyxiant may not have any toxic effects in or of itself.

Blepharospasm The involuntary closing of the eye that occurs when the eye is irritated and/or injured.

Bronchospasm A spasmodic contraction of the bronchioles due to the inhalation of an irritant.

Capsicum A lacrimating agent made from hot peppers; "pepper gas."

Carboxyhemoglobin Hemoglobin bound with carbon monoxide.

Cell lysis The destruction of cells.

Chemical asphyxian A chemical that acts in the body to interfere with the transportation of oxygen or hamper its use on the cellular level.

Chemically induced pulmonary edema, noncardiogenic pulmonary edema Pulmonary edema stimulated because of injured lung tissue, not increased pulmonary blood pressure.

Chronic Refers to a long time or of long duration. Can be used to explain either long-term exposure or long or late onset of symptoms.

Contamination The process whereby a person or piece of equipment has contact with a toxin or hazardous substance.

Cold zone Also called the safe zone or support zone the area where equipment and personnel directly support the incident.

Cytochrome oxidase An enzyme responsible for the movement of oxygen within the cell during cellular metabolism.

Decontamination A process by which the contaminants are washed from the patient, victim, rescuer, and equipment; decon.

Dirty bomb A conventional bomb that has chemical, biological, or radiological agents. The bomb disperses these agents over a large area.

Dispersal system A mechanism or process by which a contaminate is scattered within a given area.

Facciculations Uncontrolled muscle tremors.

Gross decon A process within the decontamination corridor by which all clothes are removed from the victim.

Hemoglobin A molecule found in the blood that is responsible for transporting oxygen from the lungs to the cells.

Hot zone The zone immediately surrounding the chemical or biological release. This zone extends far enough to prevent adverse effects to personnel.

Hydrolysizes A chemical reaction that decomposes, by which a compound is resolved into other compounds.

Interference gas Gases similar to those being tested that an instrument may misinterpret.

Ionization potential (IP) The minimum energy required to release an electron or a photon from a molecule.

Irritant A chemical that causes inflammation to tissues

Methemoglobin Hemoglobin that has had the iron atom converted from ferrous to ferric iron and is unable to carry oxygen.

Neurotoxin A toxin that affects the nervous system as the target organ.

Persistence Because of the mixture of the chemicals this substance can exist in the environment for a long period of time (thickened agent). It is a long-lasting effect that is desired.

Personal protective equipment (PPE) The equipment used to protect a worker from the effects of a chemical or biological

Political Barriers Obstacles to achieving one's political goals, such as money, acquiring educated individuals, and physical space.

Positive end expiratory pressure (PEEP) A positive pressure at the end of the expiratory cycle when the inthoracic pressure is normally equal to the ambient pressure.

Preplanning The act of identifying target hazards within a response area, forecasting possible emergency situations and the actions that should occur at such an event.

Radio-free area An area that has been identified as the critical hot zone in which any radio transmissions may trigger an explosive device. Hand signals should be used as a communication link to the cold zone.

Sarin Organophosphate military nerve agent.

Secondary blast The wave of energy produced by a secondary device. Can be used to disperse warfare agents after the initial event.

Secondary contamination Contamination from a previously contaminated person, liquid, or object that occurs away from the initial scene or that can be brought in from the exclusion zone.

Secondary decon The second step in the decontamination procedure. Usually consists of wet decontamination. See Wet decon.

Secondary device A device that is placed in order to maim or destroy first-response personnel. It detonates after the main charge has exploded and first-response personnel have gathered on the scene.

Secondary wave A wave created when the air around the blast rushes in to fill the vacuum created during the original explosion.

Self-contained breathing apparatus (SCBA) A closed system for breathing that contains a limited supply of air for use in a hazardous atmospheres.

Shuffle pit A corridor within the contamination reduction zone that uses chlorinated lime powder or technical decontamination solution to destroy contamination carried on the feet or shoes prior to leaving the decon line.

Soman Organophosphate military nerve agent.

Supplied air respirator An air mask unit using air hoses from a large resource of air to supply air to the user.

Tabun Organophosphate military nerve agent.

Technological barriers Obstacles to producing a weapon, such as materials, educated individuals, and weapon design.

Thickening agent A substance that is added to a chemical warfare agent in order to give a lasting effect. See Persistence.

Warm zone A buffer area surrounding the hot zone where decontamination occurs.

Wet decon The process of removing a hazardous substance from a person or piece of equipment. A process by which the contaminants are washed from the patient, victim, rescuer, and equipment, using water, soap, or technical decon solutions.

INDEX